W9-BPQ-719

FORGETTING

FORGETTING

THE BENEFITS OF NOT REMEMBERING

SCOTT A. SMALL

CROWN
NEW YORK

Published in the United States by Crown, an imprint of Random House,
a division of Penguin Random House LLC, New York.

CROWN and the Crown colophon are registered trademarks
of Penguin Random House LLC.

The cover art is from Louise Bourgeois's collection *Ode à l'Oubli* (Ode to Forgetting), a fabric book of art Bourgeois constructed from sixty-year-old monogrammed hand towels she discovered from her 1938 wedding.

All illustrations by Nicoletta Barolini/Columbia University.

Library of Congress Cataloging-in-Publication Data
Names: Small, Scott A., author.
Title: Forgetting / Scott A. Small.
Description: New York: Crown, 2021. | Includes index.
Identifiers: LCCN 2021011986 (print) | LCCN 2021011987 (ebook) |
ISBN 9780593136195 (hardcover) | ISBN 9780593136201 (ebook)
Subjects: LCSH: Memory disorders. | Memory. | Cognition.
Classification: LCC BF376 .S56 2021 (print) | LCC BF376 (ebook) |
DDC 153.1/2—dc23
LC record available at https://lccn.loc.gov/2021011986

Printed in the United States of America on acid-free paper

crownpublishing.com

2 4 6 8 9 7 5 3 1

First Edition

Book design by Jo Anne Metsch

To Michelle Small,
in memoriam

To Alexis England,
for a lifetime of memories

When I complain of the [memory] defect of mine, they do not believe me, and reprove me, as though I accused myself for a fool. . . . But they do me wrong; for experience, rather, daily shows us, on the contrary, that a strong memory is commonly coupled with infirm judgment.

It is not without good reason said "that he who has not a good memory should never take upon him the trade of lying."

—MICHEL DE MONTAIGNE, *Essays*, 1572

CONTENTS

FORGETTING

PROLOGUE

> Funes not only remembered every leaf on every tree of every wood, but even every one of the times he had perceived or imagined it.
>
> I suspect, nevertheless, that he was not very capable of thought. To think is to forget a difference, to generalize, to abstract.
>
> —JORGE LUIS BORGES, "Funes, the Memorious"

As a memory specialist, all I hear about is forgetting. And not just from my patients, who typically have disorders that cause pathological forgetting and are expressing a valid medical concern. I mean from nearly everyone else, the vast majority of whom are complaining about normal forgetting, the forgetting that we are born with and that naturally varies among us like height or other traits. I am not complaining about these complaints. My own forgetting is frustrating, and besides, offering compassionate advice is a privilege of doctoring. I am quite sure, in fact, that my early interest in memory—triggering my academic interests, training, and career—was influenced by my own forgetting. Who wouldn't want to have a better memory? To perform better on exams; to remember with high fidelity books read or movies watched; to have more details at the tip of one's tongue to win over minds in intellectual debates or hearts with fun facts and poetry?

That forgetting represents a glitch in our memory systems, or a nuisance at the very least, has always been the common scientific view. Accordingly, science has primarily focused on figuring out how the brain forms, stores, and retrieves memories; how the memory snapshot is taken, processed, and cataloged. While some scientists have intuited forgetting's possible benefits, the fading of memories, like musty photos in an attic, is usually considered a malfunction of the recording device or a sign of the flimsiness of the record. This standard view, in which better memory is always the noble goal, whereas forgetting is to be prevented and fought tooth and nail, has guided my training and my career.

I have been a student of memory for more than thirty-five years. As an undergraduate in experimental psychology at NYU, I wrote my first published paper and graduating thesis on the topic of how our emotions can bias what we see and how we remember. As an M.D./Ph.D. student at Columbia University, I worked in the lab of the memory researcher Eric Kandel, who in 2000 was awarded the Nobel Prize in Physiology or Medicine for his discoveries regarding how neurons in different animal models remember. And since completing a postdoctoral fellowship at Columbia on Alzheimer's disease and other forms of memory failure with Richard Mayeux, a leading Alzheimer's clinician and geneticist, I have been toiling away in my own lab investigating the causes of, and potentially the cures for, Alzheimer's disease and other reasons for memory failure in later life.

While old dogs can't learn new tricks, it's a good thing we can forget old ones, since it turns out that I, along with many other memory investigators and memory doctors, was wrong about forgetting. Recent research in neurobiology, psychology, medicine, and computer science has contributed to a clear shift in our understanding. We now know that forgetting is not just normal but beneficial to our cognitive and creative abilities, to our emotional well-being, and even to societal health.

This book is dedicated to the hundreds of patients I have tried to help throughout my career who have suffered from pathological forgetting, commonly caused by neurodegenerative disorders but also simply by aging itself. While the medical definition of "pathological" is often up for grabs, the easiest distinction between normal and pathological forgetting is that the latter reflects a true worsening of one's memory, a worsening that impacts the ability to fully engage in our information-laden lives. Only by observing the painful consequences of pathological forgetting in patients does normal forgetting emerge in sharp relief. Witnessing the suffering caused by Alzheimer's disease inoculates against the temptation to poeticize it—to say, for example, that the pathology carries a silver lining, that it is somehow good. Perhaps. But as a doctor who tries to sympathize as much as possible and is exposed to the suffering caused by pathological forgetting, I cannot abide that position. In any case, this book is not about that. It is about normal forgetting.

The question presented earlier—"Who wouldn't want to have a better memory?"—was obviously posed rhetorically.

What about photographic memory, a memory system that immortalizes like a computer hard drive, where snapshots never fade in a mind that never forgets? Most of us have fantasized about this cognitive power, but I suspect many have also sensed its potential burdens. Despite occasional claims in the annals of neurology, cases of true photographic, sometimes called eidetic, memory turn out to be vanishingly rare. There are some people whose natural memory is at the superior end of the normal distribution, like extremely tall variations in height. There are experts in certain fields who depend on particular memory skills so extraordinary as to seem out of this world: memory specifically for chessboard arrangements by grand masters, sheet music by concert pianists, limb movements by professional tennis players. There are also so-called mnemonists, or memory magicians, who rely on cognitive tricks of the trade, some innate skills, and lots of practice to develop superior memories for specific information categories—autobiographical information, numbers, names, or events. When formally tested, however, none of them, it turns out, have true photographic memories for all things. None have minds that never forget.

So photographic memory is really a fabrication, a superhero power. Is it desirable? Until science caught up with fiction to show why not, fictional descriptions provided the early answers. The best example is Jorge Luis Borges's short story "Funes, the Memorious," found in his anthology *Ficciones.* After being thrown from a horse and knocked unconscious, Funes awakes with an inflamed brain that can no longer forget. Now, in a flash, he can memorize and recollect anything and everything. Envy is what most readers experience when starting the story, as we learn that Funes, endowed with his new super cognitive powers, can effortlessly recite long passages from books recently read or master new languages (even Latin!) in a matter of days. But jealousy transitions to compassion as we begin to appreciate his psychic chaos. In one instance, when Funes is offered a glass of wine produced in a neighbor's vineyard, his mind drowns in a deluge of memories. The wine revivifies so many associated memories, each with pointillist details—for example, the "shoots, clusters, and grapes of the vine" from which the wine was pressed—that this total recall seizes Funes with anxiety. No wistful and meandering remembrance of things past for poor, suffering Funes. When asked about any previous event, even a lovely childhood afternoon, his mind is overloaded with the day's minutiae—the formation of every cloud seen, the minute-by-minute fluctuations in temperature felt, the choreography of every movement of his limbs. Very quickly we realize that total recall can be a nightmare.

What's most remarkable about "Funes, the Memorious" is how astutely it foreshadows neuroscientific research into the way a mind that snaps and stores photographs with retina display resolution can impair our thinking. Many of the story's passages about Funes describe a dominant cognitive impairment caused by his photographic memory: the inability to generalize—to see the forest for the trees. "His own face in the mirror, his own hands, surprised him on every occasion. . . . It was not only difficult for him to understand that the generic term *dog* embraced so many unlike specimens of differing sizes and different forms; he was disturbed by the fact that a dog at three-fourteen (seen in profile) should have the same name as the dog at three-fifteen (seen from the front)." Having a photographic memory was so distressing that young Funes wound up spending the rest of his life sequestered in a deliberately darkened and pitch-quiet room.

Only in the past decade or so has a new science begun to coalesce to explain why forgetting in balance with memory is the true and very natural cognitive power bestowed on us to live in an ever-fluctuating world, one that is also often frightening and painful. The "right to be forgotten" was legally determined in European courts in 2010, when it was successfully argued that a permanent record, in this case on the internet, could have damaging consequences for a person's life. In a similar spirit, our brains are right to forget.

As you shall see in this book, forgetting in balance with memory is required for sculpting cognition: for granting us the flexibility to accommodate to an ever-changing environment, for extracting abstract concepts from a scattered mess of stored information, for seeing the forest for the trees. Forgetting is required for emotional well-being: for letting go of resentments, neurotic fears, and hurtful experiences that fester. Too much memory or too little forgetting imprisons with pain. Forgetting is required for societal health and for creativity, lightening the mind for those eureka moments when unexpected associations are made. Without forgetting, all flights of creative fancy would be moored by memory.

If the rhetorical question posed at the beginning was changed to, "Who would want to have a photographic memory with a mind that never forgets?" I hope that after reading this book, you will appreciate that the answer is nobody.

ONE

TO REMEMBER, TO FORGET

"My mind was like a steel trap!" declared Karl, my first patient of the day at Columbia University's Memory Disorders Center. Among the many metaphors for memory, a steel trap is one of my least favorites—partly on aesthetic grounds (the violent visual of a trapped paw is repugnant), but mainly for its misleading scientific implications. Even for those in the superior range, memory is never steely; it is flexible, form-shifting, and fragmented. The trap metaphor is also mechanically incorrect by implying that memories are formed instantly, with a decisive snap.

A Manhattan criminal defense lawyer, Karl appeared dressed for court. Renowned for its expertise in Alzheimer's disease and related disorders, our center provides care to a diversity of patients from around the world. Still, Karl stood out, and not just because of the tailored fit of his three-piece suit. When I arrived punctually to the office from my research laboratory a block away, Karl was pacing and raring to go, revealing a hyperkinetic eagerness atypical of our patients. It turned out that he had studied English literature as an undergraduate at Yale, and once he'd dispensed some initial glib proclamations about his superior cognitive abilities and courtroom prowess, he relaxed and began articulating his symptoms and his fears about their causes and their consequences on his active legal career.

Listening carefully to a patient's symptoms and clinical history is the first order of a neurologist's business. These testimonials contain rich information we need to achieve our central goal, "localizing the lesion." Even more than other medical specialists, neurologists are typically obsessed with asking "where?" before asking "what?" The cause of arm weakness, for example, can localize to muscles or nerves, or different parts of the spinal cord, or the brain, and each part of this nervous system map is targeted by different diseases. Most of us admit to taking joy in solving this anatomical puzzle. It requires knowledge of the nervous system's circuitry, an understanding of how different nodes of the circuit function, and, accordingly, how to probe the circuit in isolating the source of the problem. The joys of our profession aside, localizing the lesion—answering the question of where—is critical for getting the diagnosis right.

Isolating the anatomical source of someone's malfunctioning memory is harder than isolating the source of a malfunctioning arm, but the same principle applies. Memory specialists begin localizing the lesion that is causing pathological forgetting the second a patient walks through the door. Even during the informal introductory phase, we try to map the function of a patient's cognitive brain regions in order to get a sense of how their memory network functioned in their "premorbid state"—that is, before the onset of cognitive symptoms. (Beware, an occupational hazard: we reflexively perform these functional "biopsies" even when chitchatting in social settings. Just listening to how someone tells a story—that is, how embroidered the details, how rich the vocabulary and syntax—we can't help but start color-coding the storyteller's cognitive brain regions according to level of function.) These admittedly blurry cognitive brain maps are a helpful starting point from which we can chart the anatomical source of a patient's chief cognitive complaints. By the end of the first visit, we try to formulate an opinion on the "where" of memory loss. Subsequent clinical tests—which may include blood work, neuroimaging studies like MRI, and a neuropsychological evaluation—ultimately confirm or help us adjust that opinion.

Karl had always excelled in school, and even among his academically competitive peers his memory was exceptional, as evidenced by his ability to memorize baseball stats as a boy growing up on Long Island, poems as an undergraduate, and legal torts as a student at NYU's law school. His outstanding memory was professionally useful and well known throughout his law firm. After meeting someone once—a summer intern, a legal assistant, or, of course, a client—he would never forget that person's face or name. And this turned out to be his chief clinical complaint: a slippage in recalling clients' names. Recently, a few months after meeting an important new client, he'd run into her on a bustling Manhattan street and, shockingly, fumbled for her name.

Simply an embarrassing nuisance for most of us, stumbling over a name felt to Karl like a serious drag on his career.

Just hearing Karl's history, and listening carefully to his chief clinical complaint, I began formulating a pretty good idea about which parts of his brain might be the source of his pathological forgetting. In fact, I had a strong hunch that it was likely localized to one of two areas, a hunch that I would then try to confirm with a neurological exam and a rudimentary memory test that I perform in my office, then with additional tests that would be administered at the visit's end. But in order to explain my hunch—and before we can begin to explain how forgetting works—I think an overview of memory will be helpful as I take you through my clinical thinking and evaluation, and on to Karl's ultimate diagnosis.

Among the many metaphors for memory, a personal computer is a good one. In fact, better than a metaphor, how a personal computer works turns out to be an excellent analogy for how our brains store, save, and retrieve memories. That's no coincidence, since both computer and brain engineers have to solve the same three problems in learning how best to handle vast amounts of information: where to store memories, how to save memories in the dedicated storage site, and how to open and retrieve memories on demand. For this play of memory, our brains have three main anatomical actors. A collection of regions toward the back of the brain, which for simplicity I will call the posterior area, is where many of our most cherished memories are stored. A structure buried deep in the brain's temporal cortex, the hip-

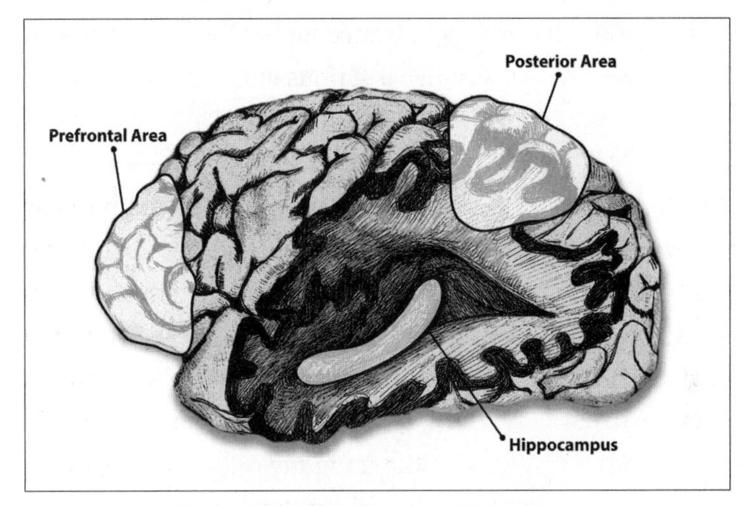

Brain areas of memory and forgetting.

pocampus, allows the brain to properly save these memories. And an area in the prefrontal cortex, located right behind our foreheads, is the general area that helps us open and retrieve memories. Whenever you save a document onto your computer hard drive or open a previously stored file, you are playing with your computer's memory just as your brain does with your own.

The same way the basic units of information storage in your computer hard drive are "bits" (the binary digits 0 and 1), the basic unit of memory storage within our brains is a cell, a neuron. But not the whole neuron; memory bits are located at the neuron's tips. Just looking at a neuron, you can see that most of it is made up of branch-like extensions called dendrites. At the very tip of the outer dendrites are tiny protrusions—hundreds of them—which are called dendritic spines. Tiny but mighty, and like budding leaves on the branches of a highly arborized tree, spines are where neurons connect and communicate with one another at a meeting point called a synapse. The larger the spines, the stronger the synaptic connection, and thus the louder and clearer the communication. As different as the neuron might appear from other cells in our bodies—say, ovoid liver cells or cuboid heart cells—it is the connecting synapse, the cleft between communicating neurons, that truly distinguishes it. If the function of an organ can be simply defined by the one distinguishing property of what its cells do—liver cells detoxify, heart cells pump—then "brain cells synaptically connect" is a pretty good definition of brain function.

Because the size of dendritic spines constantly morphs with experience, synaptic connections are said to be plastic rather than stapled with steel. When a neuron and its neighboring neuron are simultaneously stimulated at a high enough level, their spines may grow. When a sufficient number of spines proliferate, the connection between the neurons is strengthened, which is what happens when a new memory is formed. This binding of neurons explains the scientific rhyme "Neurons that fire together wire together." When the neuron is stimulated out of sync with its neighbor, the spines might wilt back down—which is what happens during forgetting. The dendritic spines, at the outer tips of neurons, are thus the information bits of our memories.

So fundamental is the shape and size of the spines for brain function that they contain a collection of molecular tools entirely dedicated to the delicate process of their growth. All neurons across the brain have these spine-growing tools, but for simplicity's sake I will call the ones found in memory-related neurons the "memory toolbox." A

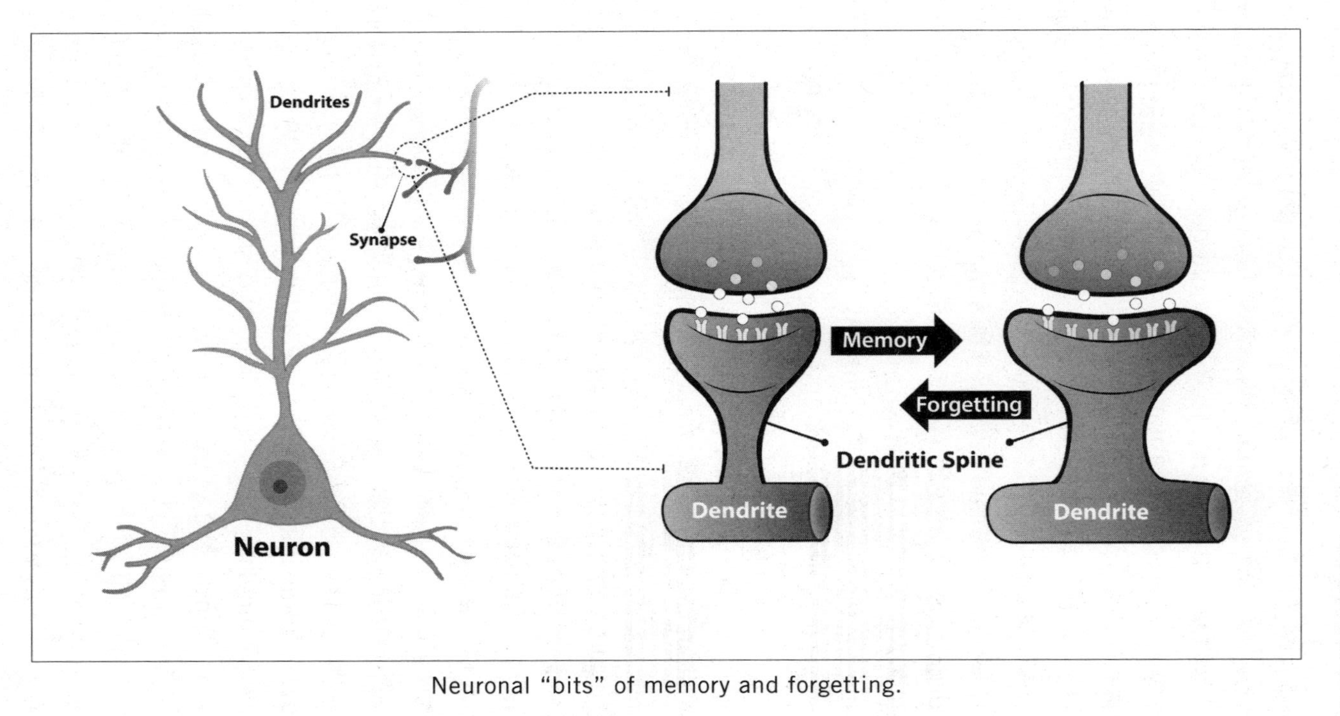

Neuronal "bits" of memory and forgetting.

lot of energy is required to grow spines, and more than that, neurons need to sprout spines carefully. If overexuberant growth were to happen, neurons would become bushy with spines, and their communications would become staticky or worse. An overgrowth of spines could cause the volume with which neurons communicate to be turned up way too high, resulting in something equivalent to screeching and uninterpretable screams. To walk this fine line, the memory toolbox grows spines with energy-saving efficiency, and does so in a carefully calibrated way.

Karl's ability to form a memory connecting his client's face and name did not occur between two neurons. Rather, the memory occurred when the millions of neurons that coded for her face strengthened their connections with the millions of neurons that coded for her name. Forming associations between separate stimuli is at the heart of memory. I'm sure you can think of many of your own complex memories that have bound multiple sensory elements into one. Karl's learning to associate a new client's name with their face can be considered exhibit A of associative memory, and for neuroscientists face-name binding has emerged as a favored paradigm for use in the lab. Ostensibly, this binding is just between two simple component parts. But with faces seen and names typically heard, the binding must bridge two separate sensory modalities, each of which is processed in different areas of our brains.

Moreover, while in our mind's eye a face might seem like a singular whole, for us to experience this singularity the brain must first reconstruct its multiple component parts—e.g., the shape of each facial feature and how the features are spatially organized. When we hear a name, a different part of the brain must similarly reconstruct the whole from individual auditory components. Face-name binding, therefore—so commonly experienced, seemingly automatic and so deceptively simple—contains within it the full complexity of associative memory.

Luckily, face-name binding is relatively easy to play with in the lab. We can generate a deck of faces and a recording of names, then use these stimuli in different combinations, durations, and sequences, thereby investigating each stage of the process that gives rise to an associative memory. Face-name binding has become the experimental equivalent of butterfly mounts, pinning down an otherwise flashing dynamic process, observing up close and in detail memory's fluttering complexity. For these reasons, many laboratories, including my own, have devised experimental protocols using MRI scanners to map brain activity while subjects view faces or hear names, alone or in pairs, and when one item is used in an attempt to have the subject recall the other. Karl's chief complaint could, therefore, be experimentally observed and investigated.

Based on these studies, I can tell you what happened in Karl's visual cortex, located in the back of the brain, when he initially met his new client. The brain first breaks down any complex item into its component parts, and then dedicated areas of the cortex reconstruct the whole. This process of reconstruction follows a "hub-and-spoke" design, similar to the way large airlines have regional hubs that converge on a central hub. The visual cortex, which handles visual information, starts with the equivalent of regional hubs that reconstruct individual facial features from basic visual elements like color and shape. These lower-level hubs then converge on higher-level ones, and ultimately on the final central hub, where the unified whole—in Karl's case, his client's face—is put back together.

In parallel, when Karl first heard his client's name, his auditory cortex performed a hub-and-spoke reconstruction of the auditory elements, ultimately representing her name at a central hub in this cortex.

I can use MRI scanners to pinpoint the precise anatomical coordinates of each level of Karl's reconstructions of his client's face and name. If I were to ask my neurosurgery colleagues to place an electrode at each level, using the electrodes to stimulate the lower hubs would not reactivate the experience of seeing a face or hearing a name. Only when the neurons in the central hubs were electrically stimulated would Karl reactivate the experience of seeing his client's face and hearing her name.

The central hubs cluster together in the posterior area, where memories are ultimately stored. In close proximity, the neurons of the central hubs are synaptically connected with one another. When synchronized stimulation occurs with high enough intensity, the memory toolboxes in the neurons open up and the tools are activated. Once the new spine growth is complete, the face and name hubs are bound together. Now, in Karl's case, the next time he ran into his client, his name neurons would be activated and he would recall her name—or, at least, that is what happened with cognitive ease in Karl's younger mind.

Lesions anywhere across Karl's visual and auditory corti-

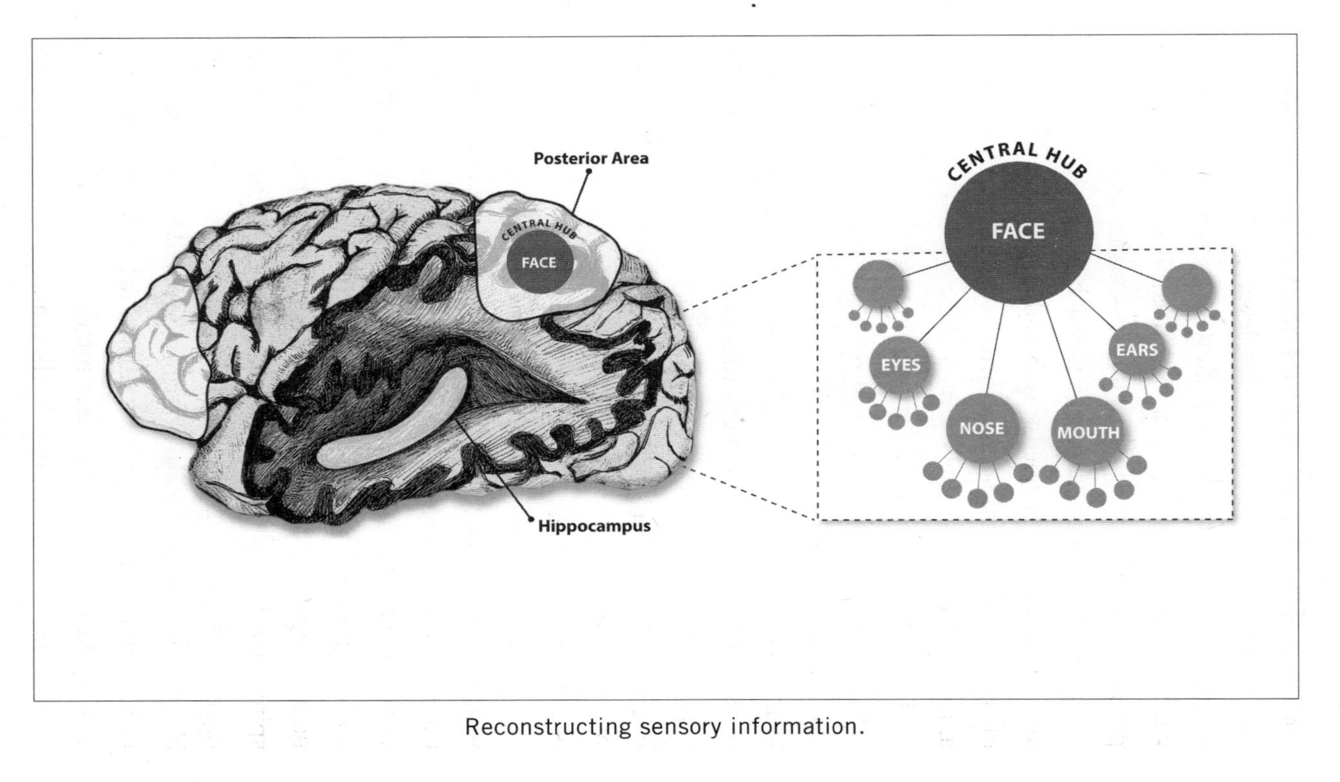

Reconstructing sensory information.

ces might have potentially impaired his ability to remember names. A stroke or tumor at the lowest hubs, for example, could block the flow to the higher hubs, and thus block the reconstruction of faces or names. But this stemming of sensory flow would do more than prevent the sensory stimuli from reaching their central hubs; it would cause what we call cortical blindness or deafness. Rare neurodegenerative diseases can cause mid-level hubs to malfunction and die, which similarly prevents the reconstruction of faces or names and leads to patients' experiencing disjointed perceptions. There are even cases where circumscribed lesions localize to the highest central hub. If the lesion affects the face central hub, the patient will have difficulty recognizing faces—not just specific faces, but faces in general, a disorder we call prosopagnosia. (For more on this fascinating condition, see the masterwork by my much-missed former colleague, the brilliant neurologist Oliver Sacks.)

Karl's neurological exam seemed to preclude lesions in his visual or auditory cortex. Pending an MRI that I would order to further exclude them, in my initial evaluation I needed to start entertaining other anatomical sources of his forgetting. When I asked Karl to elaborate on his symptoms, he returned to that episode of seeing his new client on a bustling Manhattan street a few months after they first met. Her name ultimately came to him, he said, but only after more time and with more mental work than in the past, when names—or baseball stats or poems or legal torts—would just "pop" into his mind with ease. This told me that the connections between the face and name central hubs still existed in Karl's brain, but they were working less efficiently than before.

Karl's description of memories as typically "popping" into his mind was telling. Many forms of memory work subconsciously—for example, my memory of learned motor skills that just allowed me to type this sentence—and all forms of memory use the same mechanisms of synaptic plasticity to strengthen connections among neurons. But memories that we can consciously recollect, called "explicit memories," are formed by associations among many central hubs. When Karl first met his client, he didn't simply see her face and hear her name. The episode also contained all the trappings of the place he first met her—that is, his office—and the time of day the meeting occurred. It might have included other sensations too, like the scent of her perfume. Each component was reconstructed in a different central hub of Karl's cortex, all converging on the posterior memory storage area, all strengthening those connections when experienced simultaneously in time. The stronger those components snap together, the stronger the explicit memory of his client pops into his conscious mind.

The strength of this snap depends on a separate structure of the brain, the hippocampus. We have two hippocampi—curved, cylindrical, pinky-sized structures nestled deep along the base of our temporal lobes. Sixteenth-century anatomists often used their Renaissance-infused imaginations to label newly identified brain structures, and to them the C-shaped hippocampus looked like a seahorse. Because of its sculptural elegance, anatomists long wondered what the hippocampus did, but its function remained a mystery until the 1950s, when in an attempt to control debilitating seizures in a twenty-four-year-old with intractable epilepsy, neurosurgeons decided to remove parts of his brain, primarily both of his hippocampi. The seizures stopped, but after the operation the patient was unable to form new conscious memories. His conscious memories up to a few months prior to the surgery were largely intact, as was his ability to learn subconscious memories (e.g., learn to perform new motor tasks).

When introduced to a new doctor, he would be able to converse with her using her correct name as long as she stayed in his room. The hubs and spokes of his sensory cortex functioned normally. But if the doctor left the room for a few minutes and then returned, not only could the patient not remember her name, but he had no conscious recollection of ever having met her. This severe form of pathological forgetting occurred even when seeing the same doctor, who tended to his care for decades, over and over again. He could never again associate a face with a name or link cortical hubs of any kind. Never again would any new context, event, or place pop into his mind as a conscious memory. He lived for more than fifty years after his surgery, apparently without any psychic suffering over a condition that is clearly devastating. He simply had no conscious memory of his forgetting.

Now that the patient has died, we can use his name publicly, Henry Molaison. His initials are still used when describing him in the literature, and decades later his case study is still revisited. Although the surgery was performed with good intent, crippling H.M.'s cognition is not the field's proudest moment. Nevertheless, his legacy lives on, having ushered in a new era in cognitive science in general and the study of memory in particular. Decades and thousands of studies later, we now have a better, if not complete, understanding of how the hippocampus helps us form new conscious memories.

The binding of information across the central hubs turns out to be a slow and deliberate process. The dendritic spines of their neurons, those that bind the hubs, proliferate when stimulated, but this proliferation is particularly fragile and unstable. Without continuous stimulation, they tend to shrink back down. Like inattentive first graders, neurons in the central hubs are considered the "slow learners" of memories. The hippocampus overcomes this synaptic fidgetiness by functioning as a compassionate but strict schoolteacher who patiently teaches these unruly new spines to stabilize and the hubs to bind into a conscious memory. Once the central hubs are educated and the posterior area has stored a conscious memory, the teacher—the hippocampus—is no longer needed and can move on.

Two important discoveries have exposed the secrets of the hippocampus's cortical training program. One is that the hippocampus has direct lines of communication with each of the central hubs in the cortex, functioning like an old-fashioned telephone switchboard. The collection of central hubs that define the components of an episode—its sensory elements, the time of day, the place—ring into the hippocampus, with each hub stimulating different patches of hippocampal neurons. The second discovery is that hippocampal neurons grow dendritic spines differently than cortical hub neurons. Hippocampal neurons are fast learners, and upon co-stimulation they quickly grow into new, fully stable, mature spines.

When Karl first met his client, the central hubs that defined that introductory episode—representing her face, her name, and all the other elements of the moment—simultaneously rang into his hippocampi, with each hub co-stimulating different patches of hippocampal neurons. If the hippocampal patches were keys on a piano keyboard, then meeting the client struck in Karl's hippocampi a chord of many memory notes. Since hippocampal neurons bind quickly, for a short time after such an episode the hippocampus acts as an indirect intermediary binding the central hubs. Once the hippocampus has the focused attention of all the central hubs that initiated that first meeting, it can begin to teach by co-stimulating each hub. Gradually, over the course of weeks, the dendritic spines in the hubs overcome their natural resistance to learning and forming stable spines. At this point, the conscious memory is said to be hippocampal independent, which is a good thing because hippocampal neurons are as fast to deconstruct their new spines as they are to construct them. It's as if the hippocampus, once discharged from its pedagogic duty, just shreds the course material, deleting it from its records.

This explains why H.M. did not need his hippocampi to recall older memories: his cortex was intact, and all those older memories had already completed their hippocampal training program. But once his hippocampi were removed, his brain was incapable of learning any new conscious mem-

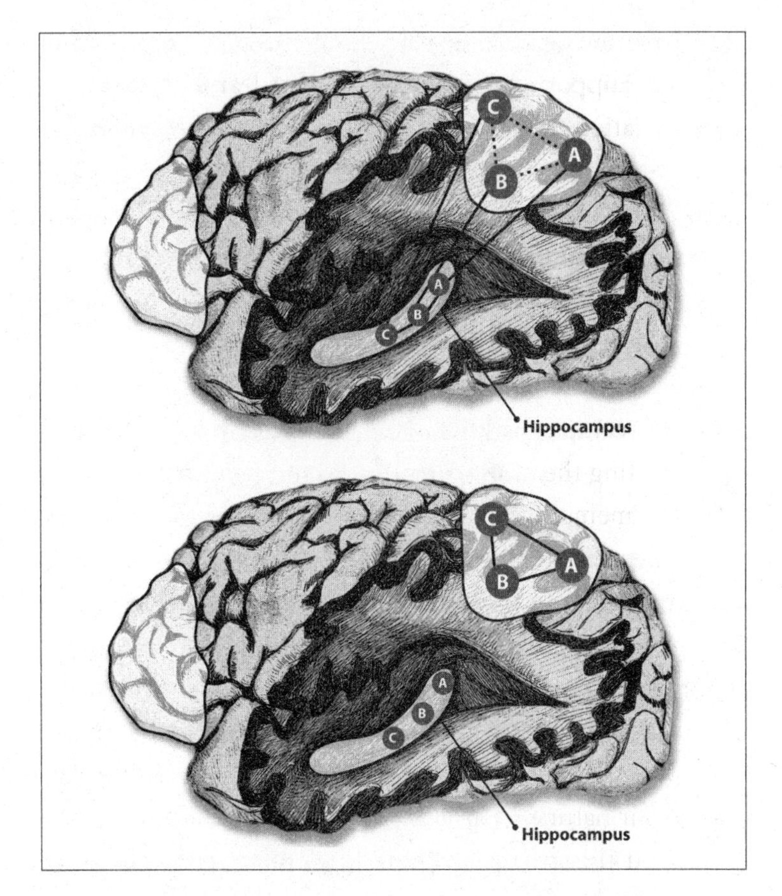

The hippocampus and memory formation: on the top is during hippocampal training, and below it is after.

ories. To use professional terminology, he had "anterograde amnesia." He also had some "retrograde amnesia" (the inability to remember old memories) for episodes that occurred a few months before his surgery. This window of retrograde amnesia was on a time gradient, with episodes that occurred in the past few weeks completely forgotten, and those that occurred in the past few months forgotten less. True retrograde amnesia, the dense kind without any time gradient, happens when the memories previously stored in the posterior area are damaged—when the information within central hubs are deleted, or when the bridges connecting central hubs crumble. This kind of dense loss of all memories is neurologically a rare event that is, nevertheless, a common plot twist in soap operas.

Karl's hippocampi had to be considered as a potential source of his pathological forgetting. But so did his prefrontal area, which helps us access and retrieve our memories from their storage site in the posterior area. Think of the hippocampus as the Save button on a computer. By clicking on this function, as I just did, you transfer information from temporary to long-term memory—from your computer screen to the hard drive. If you've ever had your computer crash before you were able to save the information on your screen, you have an inkling of what H.M.'s life must have been like. As long as his doctor engaged him in her office, H.M. could remember her name and any other associated information he experienced during that episode. But as soon as his attention strayed—which would happen when his doctor had to leave the room even for a few minutes—the information was gone. Focused attention is the cognitive equivalent of your computer screen remaining on. You can remember information as long as the file containing that information is open—as long as your attention remains focused on the episode. Diverting your attention, even briefly, is equivalent to turning the computer off. If you did not begin the process of saving the information into a long-term memory storage site, which is effectively what the hippocampus begins to do, the information is lost forever.

In contrast to the posterior area, the prefrontal area is more like your computer operating system's Open button. By clicking Open, you can scroll through your saved files, retrieve the right one, and recall it to your computer screen. Similarly, the prefrontal cortex scrolls through and recalls the memories saved in the cortical storage area.

The computer analogy is a good one, but like all analogies, it is imperfect. Unlike saving a document on a computer, saving a memory in the hippocampus takes weeks, which helps explain why the ability to memorize new conscious memories varies among people. For some, and certainly for Karl when he was younger, the hippocampus operates more efficiently than for others. And whereas you cannot open *any* files on your computer without an Open command—the computer on its own has effectively no memory retrieval—patients who, because of rare diseases or traumatic accidents, have little to no prefrontal area function still maintain the ability to retrieve and recall previously stored memories. They simply do so more slowly and less accurately.

Returning to the teaching metaphor, if the hippocampus is analogous to a schoolteacher, then the prefrontal area is more like a school librarian. Retrieving a memory is like retrieving a book from your school library. You could probably manage to find the book you were looking for without the help of a librarian, but a good librarian could expedite the process.

. . .

As Karl's first visit was winding down, I suspected that his forgetting was caused by dysfunction in one of two regions: his memory teacher, the hippocampus; or his memory librarian, the prefrontal cortex. In theory, diminished function in either region could result in the same memory complaint. There are a handful of tricks of the trade that could help me dissociate these two regions in Karl's case. I asked him whether forgetting names occurred only for clients he had recently met or also for those he had met years ago. If the prefrontal cortex was the source of forgetting, it should have been equally difficult to retrieve both new and old memories. If his complaint was just remembering names of new clients, it would implicate the hippocampus.

I also asked Karl about any difficulty finding words. The bulk of our native vocabularies have been memorized in our cortex by the time we are young adults. Most of us have experienced the frustration of failing to retrieve that one perfect word—mid-sentence, mid-story—only to have the recall gun go off later. When these episodes of verbal blanks occur with increasing frequency, neurologists immediately consider a problem in the prefrontal cortex.

I asked Karl about spatial forgetfulness as well. Not only does the hippocampus bind the sensory components of an episode, but it is also distinctly good at binding the spatial components—say, the office where Karl first met his client. I often ask patients if they tend to forget where they parked their car, how to navigate familiar driving routes, or where they put their keys. When you walk out of a shopping center and know precisely where your car is, or when you leave for work in the morning and remember exactly where your keys are, you experience your hippocampi at work. If these sorts of spatial memories become more difficult over time, your hippocampi are telling you—and alerting me, your neurologist—that their function is starting to change.

Karl was adamant: he never forgot the names of his older clients and, proud of his extensive vocabulary, had no difficulty finding words. But he did admit to rare instances of spatial forgetting: feeling uncertain about where he'd parked his car after coming out of the mall with his grandson, for instance. He'd always blamed the size and bustle of the mall for his lapse, but in retrospect he noted that his twenty-three-year-old grandson had no problem remembering where the car was parked.

At the end of the initial evaluation, I felt fairly confident that I had localized Karl's cognitive problem. For all his life, Karl's hippocampi had rung loudly, allowing the central hubs in his cortex to store new memories faster and better than many of his peers'. His forgetting localized to the hippocampi, which was still functioning, though less efficiently than when he was younger.

"Fine," Karl said as I tried to explain to him my thinking. "I congratulate you, Dr. Small, on your anatomical skills, but what's the cause?" I asked him for patience and promised that I would let him know at his follow-up visit, after we had completed a series of tests.

Diminished hippocampal function can in rare instances be caused by overt structural lesions like a stroke or tumor, or by rare hormonal or vitamin deficiencies. I excluded these possibilities with an MRI and blood tests. The most important test in cases like Karl's is a formal neuropsychological evaluation. Performed by a specialist called a neuropsychologist, this evaluation is similar to those annoying IQ tests you may have taken, consisting of a battery of pencil-and-paper or computerized tests designed to interrogate the function of your brain's cognitive regions. "Cognitive regions" include those involved with memory, but also with language, calculations, and the ability to manipulate objects in space or engage in abstract reasoning. Over the span of decades, these tests have been standardized and given to a large number of patients across a spectrum of ages, genders, levels of education, and ethnic backgrounds. The tests are as close as we can get to an objective measure of cognition, almost like an EKG for a cardiologist. In Karl's case, they confirmed that his prefrontal cortex was fine, but also that his hippocampi were subtly underperforming. Because these tests have been given to so many people, including those in Karl's age-group and at the same level of education, a neuropsychologist can estimate how well Karl's hippocampi probably functioned in his younger years. His current hippocampal function, while not worse than that of other septuagenarians, was clearly below the hippocampal performance of younger people in his demographic. Karl was experiencing hippocampal-based age-related memory decline.

There are two possible causes of gradually worsening hippocampal-based memory function in a person in his seventies: the earliest stages of Alzheimer's disease or normal aging. Alzheimer's disease begins by affecting the hippocampus, in its early stages causing mild difficulties in forming new conscious memories. Over time, the disease sweeps across other cortical regions, such as the temporal, parietal, and frontal cortices, leaving in its wake more generalized and more profound cognitive deficits, which is a defining feature of dementia. But the functional integrity of the hippocampus can also be diminished by the normal wear and tear of the aging process—the brain's equivalent of presbyopia, the normal loss of vision that occurs in all of us as we age.

Currently, we do not have precise tests that reliably distinguish between the earliest stages of Alzheimer's and normal aging, but this is about to change. Researchers are perfecting newly developed cerebrospinal fluid tests that can detect evidence of the histological brain abnormalities that typify Alzheimer's disease: protein clumps in the brain called amyloid plaques and neurofibrillary tangles. The precision of these tests is now under evaluation, but they are starting to enter clinical practice.

Another approach that can distinguish between hippocampal dysfunction caused by the earliest stages of Alzheimer's and normal aging relies on a fact about the hippocampus that has emerged over the past century, ever since neuropathologists have used brain stains that can stain individual neurons. More-modern tools have clarified that the hippocampus is comprised of several different kinds of neurons that cluster in distinct, connected regions. So, more than a singular brain structure, the hippocampus can be thought of as a brain circuit, with all of its regions acting as the circuit's nodes. The hippocampal regions are tiny islands of neurons, just a few square millimeters across, which means that in order to visualize them in living patients, we need a brain scanner that has submillimeter resolution. One of the innovations my lab is best known for is optimizing MRI brain scanners to have enhanced resolution, with the express purpose of detecting dysfunction in individual hippocampal regions. Using these tools, we have found that while the earliest stages of Alzheimer's and normal aging both affect hippocampal function, they do so by targeting different hippocampal regions. These imaging tests are currently being evaluated in hundreds of people followed for many years, the only way to learn how precisely they are able to distinguish the earliest stages of Alzheimer's disease from normal aging. We will know soon enough.

But not soon enough for Karl. Diagnosing as accurately as possible, even in the absence of a smoking gun, is a physician's obligation. During Karl's follow-up visit, with all his test results in hand, I candidly discussed both possibilities with him, and why at that point my clinical diagnosis tilted toward memory decline caused by normal aging, not Alzheimer's disease. I explained the difficulty in being sure and the current state of research. Karl's litigation skills then kicked in. He cross-examined me with a battery of rapid-fire questions, as he interrogated how I had arrived at my conclusion and established the limits of my knowledge. "What is memory?" he asked. "How is it formed? What does the hippocampus do?"

Whether you call this sort of sparring Talmudic, a debating style I first encountered in yeshivas while growing up in Israel, or Socratic, the style that forms the basis of scientific discourse, I don't mind participating in it. In fact, I like it, especially when it is engaged in smartly and with intellectual integrity. It helped that Karl's rat-a-tat-tat was peppered with a lawyer's droll humor. When hearing the case of H.M., for example, he raised an eyebrow and gleefully wondered whether the neurosurgeon was ever sued for robbing a patient of his memory-making ability.

While understandably frustrated with my uncertainty, Karl was relieved to hear that I did not think he had Alzheimer's disease. Naturally, he next asked about treatments for memory loss caused by cognitive aging. The broader question for those of us in the biomedical field is whether we should be dedicating resources to finding a cure for a condition that occurs normally as part of the aging process. Shouldn't we, the argument goes, be focusing exclusively on Alzheimer's, which is not only a true disease but also ultimately much more devastating?

I got into hot water with folks from the pharmaceutical industry when I worried aloud about this dilemma in a 1999 interview. I had been asked to appear on CNN after publishing my first papers on cognitive aging. I idly posed a provocative question to the interviewer about whether we should be developing "Viagra for the mind," which turned out to be the title of the interview. Two decades of subsequent research into the neurobiology of cognitive aging—and perhaps more importantly into its personal and societal consequences—has helped inform the answer to this question.

In 2009, I was asked to help organize a symposium on cognitive aging, which included bioethicists and representatives from the U.S. Food and Drug Administration (FDA). The emerging consensus in this and subsequent forums was that it is valid and justifiable to develop interventions to treat normal age-related memory decline, since doing so can meaningfully impact people's lives, which are now more cognitively complex and demanding than ever before. Just as it is correct and justifiable to develop reading glasses or even surgery for presbyopia, it is bioethically correct to try to correct cognitive aging.

I support this view but still believe that lifestyle interventions such as behavioral or dietary modifications are better suited for treating cognitive aging than are pharmacological drugs. Cognitive aging occurs in all of us. As life spans expand globally, cognitive aging is emerging as a worldwide epidemic. Effective lifestyle interventions, if they can be found, are better than drugs in ensuring equal access to all. Subtler in their effects on brain biology than drugs, they are also a better match for cognitive aging's less insidious pathophysiology.

My lab and others have been investigating the effects of physical exercise and dietary interventions in cognitive aging, and other labs have been investigating the effects of cognitive exercises. While there is reason to be hopeful that both dietary interventions and cognitive exercises will turn out to be useful in ameliorating cognitive aging, at this point only physical exercise meets the minimal standards for validating a clinical recommendation, so that is what I prescribed to Karl. Sensing that he would have preferred to be prescribed a drug, I braced myself for another of Karl's friendly cross-examinations, which I would have considered justified; after all, exercise is not curative of cognitive aging. But perhaps so relieved that I did not think he had Alzheimer's or appreciative that I candidly admitted my limitations as a doctor, Karl seemed satisfied. "Good," he said with wry resignation. "My wife will be pleased. She's been worrying about my weight."

I also recommended that he continue to see me for clinical follow-up. Absent diagnostic tests and medical certainty, tracking Karl's cognitive and clinical trajectory over time was the best way to confirm or refute my diagnosis. Karl decided to enroll in a number of the research studies offered at the Memory Disorders Center. The first was an observational study, funded by the National Institutes of Health (NIH), that would follow patients over time with repeated neuropsychological tests and MRIs. The second was an autopsy study at the end of life, which, as I explained, was the only way to know his true diagnosis with absolute certainty, with the added benefit that postmortem brain tissue is important for research. He jumped on this right away and declared his enthusiasm to donate his brain "for the sake of science!" in a polyphonic tone that I came to understand as typical Karl, bravado ringing ironic atop quieter worries—in this case, for his children and grandchildren. If the autopsy revealed Alzheimer's disease, he worried that it might have genetic implications for them.

Karl showed up in my office every six months. Always dressed to the nines, always prowling my waiting room, always a genuine pleasure to see. He would ask me about all sorts of memory boosters he'd read about, from supplements to meditation to yoga, often bringing in news clips as evidence. Open to anything, and humble in the face of our field's ignorance, I'd review his suggestions, often downloading and reading the primary publications that led to claims of cognitive efficacy. Some were more plausible than others, but none met the standards justifying a formal clinical recommendation. For the ones that did no harm, I suggested that he explore and let me know how it went. None worked, although surprisingly—or perhaps not—Karl ended up enjoying meditation and kept with it. Clinically, Karl was evaluated biannually till his death eleven years later of cardiac causes. Cognitively, Karl's hippocampal dysfunction—his memory loss—had worsened slightly, but it never spread to other regions of his brain, never caused profound cognitive impairment or dementia, and thus supported the diagnosis of normal, disease-free, cognitive aging.

The ultimate confirmation came from his autopsy. Once it was completed, I took two separate elevators to get down to the pathology suite located in our hospital's windowless subbasement to review Karl's brain sections with our neuropathologist. I had been studying the brain for nearly thirty years at the time, and this particular brain for eleven. But no matter how long you've studied the brain, nothing can reduce the awe you feel when faced with the postmortem brain of a person you knew. No amount of knowledge can bridge the gap between those slabs of tissue—no matter how many billions of neurons, how intricate their synaptic connections, or how ornate the architecture of their networks—and that living being.

The neuropathologist, of course, did not know Karl and didn't register the intellectual and emotional vertigo I was experiencing from this viewing of my beloved patient, his brain sections now coldly displayed in careful anatomical order on stainless steel. As if reviewing liver or kidney sections, the neuropathologist dispassionately reviewed Karl's brain, slice by slice, and then select sections under a microscope. He found no evidence of amyloid plaques or extensive neurofibrillary tangles in Karl's once fiery hippocampi or crackling cortex. No Alzheimer's disease.

Even at its peak, Karl's memory was never perfect. At one of his visits, after I'd discovered his literary bent, I gave him "Funes, the Memorious" to read. He appreciated the writing quality and got the point, but he still felt it was only a clever parable on hubris. A photographic memory is a superpower that he insisted he would find desirable. I would like to believe that the newly emerging science of forgetting, with key findings published since Karl's death more than a decade ago, would have helped me convince my argumentative friend that a photographic memory is a curse. These studies have begun clarifying the molecular mechanisms in neurons and synapses that actively govern normal forgetting. Their results stand in interesting contrast to the older and better-understood science of memory, which, through decades of research, has established that the growth of dendritic spines is memory's neuronal feature.

The first studies of the science of forgetting showed that its neuronal feature is simply memory's reverse—a shrinkage in the number or size of dendritic spines. It was plausible, therefore, to assume that forgetting was simply faulty memory, a passive rusting of memory's spine-growing tools. This type of forgetting can occur in normal aging—what Karl had—or Alzheimer's disease, both forms of pathological forgetting. But not so, it turns out, in normal forgetting. New insight in the past few years has uncovered a completely separate set of molecules involved with normal forgetting, a molecular toolbox distinct from growing spines. When this "forgetting toolbox" is opened, its tools carefully disassemble spines, shrinking their size.

Discovering that Nature granted us separate molecular toolboxes actively dedicated to memory on the one hand and forgetting on the other clearly refutes the common view that forgetting is just failing memory. But this does not necessarily mean that normally occurring forgetting is beneficial—a seductive and potentially false conclusion. After all, Nature gave us an appendix. It is possible that the forgetting toolbox serves no beneficial function and is just a vestigial holdover from some ancient period. Or, worse than a benign relic, it might be accidentally detrimental in our relatively new environments, so cognitively rich and ever changing. Perhaps once evolution catches up to these new cognitive pressures, following its own slow metronome, we will rid ourselves of the forgetting toolbox that seems to trouble us all. We will evolve to be like computer clouds, with minds of potentially infinite memory, minds that never forget.

But recent studies have shown that the molecular forgetting toolbox does, in fact, serve a beneficial purpose, conferring a clear advantage that is perfectly adapted for our complex worlds. Forgetting is a cognitive gift. Our understanding of the dynamics of synaptic plasticity, defined as changes in the strength of neuronal connections with experience, has expanded. Like a car engine or any complex dynamic system, synaptic plasticity needs both an accelerator and a brake. The shrinkage of spines is governed by the forgetting toolbox and abides by the same two rules the memory toolbox follows for spine growth—except in reverse. In contrast to spine growth, which is triggered when a neuron receives synchronized inputs, the active process of spine shrinkage occurs when the input is desynchronized in time, or when the neuron receives new inputs that override earlier ones. And just as the memory toolbox grows spines slowly but surely, the forgetting toolbox carefully shrinks spines.

The benefit of the forgetting toolbox is most clearly shown in animal models, typically flies and mice, in which the function of the specific molecules in each toolbox can be selectively manipulated and the subjects observed for what then happens to behavior. These types of molecular manipulations can't be performed in humans for obvious ethical reasons, although, as I will discuss in later chapters, accidental genetic mutations can be revealing. It's impossible to know for sure whether nonspeaking animals experience that special "pop" of a conscious memory, but I suspect that anyone with a pet would bet they do. Neurobiologists have, in any case, devised clever behavioral tasks, like neuropsychological tests in humans, to evaluate these complex memories in animals in the laboratory.

Neurons in all animals look nearly identical, including the all-important synapse, so much so that even a seasoned neurobiologist would struggle to tell the difference between pictures of a fly, a mouse, and a human neuron or synapse. We all have a similar complement of molecules in our neurons, and it is these molecules, typically proteins, that govern cellular structure and function. Not surprisingly, key molecules in the memory and forgetting toolboxes in all neurons, in all animals, are nearly identical. So, what happens when forgetting molecules are turned down, preventing normal forgetting, in animal models using an array of molecular tools? Cognitive and emotional havoc ensues. When these molecules are turned up, when forgetting is accelerated, both cognitive and emotional measures improve. More precisely, as will be illustrated throughout this book, both normal memory and normal forgetting work in unison to balance our minds so that we can healthily engage chaotic and sometimes hurtful environments.

We are students for life, and I am saddened that I did not get the chance to explain to Karl what I now better understand, what this emerging science of forgetting is starting to teach us: that his mind was blessed with the ability to forget. Not the accelerated forgetting he experienced in later life, which is pathological, but the forgetting he experienced throughout his younger years. Yes, brains with memory and no forgetting would be better at quoting baseball stats and reciting poems. Yes, brains with memory and no forgetting would excel in a never-changing world—for, say, Bill Murray's character in *Groundhog Day,* navigating fixed environments day in and day out. Those brains, like the stacks of torts in a law library, easily accessed and remembered in perpetuity, would never forget the hurt of a harm done. But, as we will see in the next chapter, a brain with memory and no forgetting would miserably fail at living all aspects of a meaningful life.

TWO

QUIET MINDS

Freddy had just started first grade and loved singing while walking to school. This was how his mother began tenderly describing her son during a visit with his pediatrician. Perhaps she was fortifying herself for the appointment's real purpose or trying to convince herself that despite some of his peculiar behaviors, nothing was really wrong. If so, the veil quickly fell. Following some other hopeful accounts—that Freddy, for example, had a remarkable memory not only for songs but also for numbers—she tearfully described how his behavior had recently turned unmanageable, disrupting life at home and school. Generally sweet, Freddy would suddenly get angry over any perceived change in his daily routine and surroundings. A slight alteration of a single volume's placement in the family's bookshelf was enough to provoke frustration and, if not immediately corrected, temper tantrums. Routines were nearly ritualized as he adamantly demanded that they had to remain the same. If his mother even considered an alternate walking route to school, a violent outburst would shatter a lovely moment of holding hands and singing together.

Freddy's pediatrician, Dr. Leo Kanner (who referred to his patient as "Fredrich"), had an active mid-twentieth-century practice at Johns Hopkins and began noticing a pattern of similar complaints he'd occasionally hear from other parents of young patients, all between four and eight years of age. Charlie, for example, would lash out if the dining room table was not set identically each night. His family was allowed to sit down to eat only after the silverware had been reset precisely as he remembered it. Susan became transfixed by and anxious about a new crack in a wall so minuscule that it had gone unnoticed by the rest of the family. Richard's inflexibility was most evident during his bedtime routine, which he insisted follow an exact, repetitive sequence of events.

Kanner, who would become known as the father of child psychiatry, collected these case studies into two seminal monographs that described a new pediatric condition. The second study, published in 1951 and titled "The Conception of Wholes and Parts in Early Infantile Autism," characterized what Kanner considered the core feature of what would come to be known as autism spectrum disorder. "The autistic child desires to live in a static world, a world in which no change is tolerated," Kanner wrote. "[The child has] an obsessive desire for the preservation of sameness. . . . The status quo must be maintained at all costs." It is likely a coincidence that around the same time that Borges wrote his neurologic science fiction about how trauma-induced photographic memory caused in the story's antihero, Funes, an obsessive desire to preserve sameness, Kanner wrote that children with autism become agitated when seeing or hearing anything that deviates from the "photographic and phonographic details" of their memories.

Most of us would walk by a familiar and fully stacked bookshelf and scarcely notice if a single volume was missing or had been transposed with another. Seeing the forest for the trees, or in this case the bookcase for the books, is what psychologists sometimes call generalization, a cognitive ability that allows us to extract general patterns from component parts, to synthesize and integrate parts into a unified whole. Kanner posited that in contrast to those with typical cognition, where the parts are easily reconstructed and seen as a whole, children with autism are overly fixated on the parts.

As in the case with Funes, Freddy and many of the children Kanner diagnosed with autism had remarkable memories, but mainly for one kind of memory, which, lacking a synthetic associative quality, is sometimes called rote memory. Memorizing a song's lyrics and tune after only one listen and reciting a long list of numbers at one go are examples of rote memory. Kanner did not clearly make the connection between memory of any kind and the ability to recognize the whole from its parts. But, illustrating how literary insight into the workings of the mind often predates scientific knowledge, Borges recognized that memory needs to be balanced by normal forgetting in order for a person to cognitively generalize. Without forgetting, young Funes could not generalize from one sensory experience to the next—did not, for example, grasp that the dog he saw in the morning light and the dog he saw at twilight were the same dog. Without forgetting, Funes found that his only respite from life's constant fluctuations was to routinize his life and to minimize sensory overload by banishing himself to his dimly lit, quiet, and never-changing bedroom.

Science has now caught up. Researchers investigating the new science of forgetting have shown that the implicit assumption underpinning Borges's science fiction was right: we depend on our normal forgetting to cognitively generalize. More than just validating Borges's assumption, science is beginning to explain why and how forgetting is required for healthy cognition.

Scientists who investigate animal models have relied on research in mice and flies to validate and explain Borges's insight. But clinical scientists investigating autism have contributed to the broader understanding of how our minds benefit from forgetting, of how forgetting helps us to cognitively engage our ever-fluctuating worlds. Many clinical scientists insist on pluralizing autism into a group of disorders, because it is assumed that autism does not have one unifying etiology. Some, including family organizations, advocate the position that autism is not even a disorder, that it is just an extreme in the normal variation of our social skills. Whether autism is multiple disorders, one disorder, or no disorder at all, recent genetic studies have identified a reliable group of genes whose function is altered in people with autism. Enough of these genes have turned out to be part of the molecular toolbox that functions in forgetting, and enough of them have been found to diminish forgetting, that they are beginning to provide a neurobiological explanation for Kanner's formulation that many people with autism desperately seek sameness in an effort to reduce anxiety-provoking cognitive chaos.

This research can help resolve what might seem like normal forgetting's greatest enigma: why forgetting—which, after all, subtracts rather than adds to our brains—is cognitively beneficial.

Assuming you woke up this morning in your own bed and not in a new home in a foreign land, many of your behaviors throughout the day will depend on how flexible your existing memories are. In fact, they rely much more heavily on that flexibility than on the memory capacity in your cortex, the number and size of dendritic spines it contains, or how well your hippocampi can fill up these cortical memory stores with new information. When yesterday's information slightly alters, say, your morning routine, your daily commute, or your interactions with colleagues at work or with your family during dinner, you're exhibiting behavioral flexibility. It is easy to imagine how we would suffer if we had an inflexible mind. No matter how routinized our lives, the continuous alterations to existing memories are vital for us to adapt to our rapidly shape-shifting worlds. Just as the most elegant home remodeling is often a combination of construction on top of demolition, the brain's optimal solution for behavioral flexibility turns out to be a balance between memory and active forgetting.

In the last chapter, we saw that neurobiology has isolated

two distinct molecular mechanisms: one for memory, the other for forgetting. Scientists can now apply experimental tools to animal models that essentially grant them access to control knobs for each pathway. With the ability to turn memory and forgetting up or down, they can then measure how different manipulations influence an animal's behavior. When an animal is learning the fastest way out of a maze, for instance, memory mechanisms obviously need to be "on" and dialed up. More memory leads to faster learning of the intricacies of the maze and its escape routes. After the animal masters the course, the maze can be slightly altered so that the animal must learn a subtly different route. Modifying the previously established memory for the original maze is more efficient than forming a brand-new memory from scratch in learning the alternate route. You might assume that cranking up memory even more would help, but in this and other examples of behavioral flexibility, the efficiency and speed of learning an alternate route actually depend more on forgetting. Turning up the forgetting control knob, while leaving the memory knob untouched, is the faster way to learn an alternate route out of a well-known maze. Behavioral flexibility is, therefore, like sculpting with marble where the forgetting chisel dominates.

Recall how, at the molecular level, mechanisms that increase memory and forgetting are equivalent for all animals, from flies to mice to humans. We all have the same molecular toolboxes that turn up memory by the growth of dendritic spines and turn up forgetting by causing spines to shrink. Still, despite the findings on the benefits of forgetting for behavioral flexibility in animal models, there is always a chance that our species is somehow special. One way to confirm that forgetting is essential for behavioral flexibility even in humans is to find individuals whose ability to forget is turned down by their genetic makeup and determine whether and how this influences their behavioral flexibility. Nature, with her broad range of diversity, has presented us this opportunity with autism.

I first met Dr. Daniel Geschwind, now one of the world's preeminent experts on autism, while completing my internship in internal medicine at UCLA in the early 1990s. At the time, Dan was in his first year of a neurology residency at UCLA, a year ahead of me on the academic path to becoming a neurologist. Although he stayed at UCLA and I moved east to Columbia to complete my training, we have maintained a lifelong friendship forged in our early careers around our shared reductionist view of the brain—believing that all behavior, no matter how complex, can be reduced to its cellular and molecular building blocks—and a shared skeptical spirit that extends beyond science, a spoofing sensibility that can, we are often told, border on the absurd.

Dan's quantitative mind served him well as a chemistry major in college and as an M.D./Ph.D. student who completed his graduate work in human genetics. While this educational pedigree granted him a deep understanding of the chemical composition and function of genes, that is not what distinguishes Dan or his research program. By the time Dan started his higher education, the "genetic revolution" was decades old and had already demystified the process whereby the code contained within each gene is "expressed" to generate proteins that govern all cellular function. Medically, this revolution led to discoveries in how mutations—glitches in individual genes—can cause rare inherited diseases like sickle cell anemia. These medical discoveries were, however, confined mainly to "simple" genetically driven disorders. Each cell has more than twenty thousand genes, and in this context "simple" means that a single genetic mutation is sufficient to cause a disease, whereas a "complex" disorder is caused by the interplay of a large number of minor genetic glitches and a range of environmental factors.

When Dan entered the genetics field, new tools were being developed to investigate the function of thousands of genes simultaneously, extending the field's reach into the molecular biology of complex disorders. Here Dan's intellectual distinction—a remarkable capacity for synthesizing and integrating a vast array of information into unified concepts—allowed him and his research program to thrive. Instead of focusing on how each individual gene functions or malfunctions, one by one, Dan was in the vanguard of a pioneering group of investigators who devised new methods to clarify how hundreds of genes function in unison, in genetic networks. Thanks to these researchers, it is now possible to ask how hundreds of genetic errors, each alone having a subtle effect, collectively conspire to cause a genetic network to malfunction in complex disorders.

A mind that can synthesize a complex mess of information might also be useful in coping with the complexity of life. I've always thought that Dan would have excelled as a life coach if, at a great cost to medicine, his medical career had faltered. While we are approximately the same age, Dan seemed to have figured out life early on, back when we were both young trainees at UCLA. He was already happily married and living in a large Spanish Colonial home in a tony part of Santa Monica, with palm trees in front and bougainvillea in its grassy backyard. Meanwhile, single and with only a suitcase to my name, I rented out a tiny room in a sandy, run-down cottage off the beach in Venice. (I did end up meeting my future wife at the end of my internship year, which is one reason why, despite my reputation as a die-hard New Yorker, I have always been soft on L.A.) Over the years, Dan and I have maintained vigorous East Coast–West Coast debates, on the quality of higher education in general and neurological education in particular, on the restaurant and arts scenes, and on L.A.'s casual sunniness versus New York's soulful seasons.

There is one debate, however, that Dan has clearly won. I have always advocated that no matter how complex a brain disorder's ultimate behavioral manifestations—dementia in Alzheimer's disease, mental derangements in schizophrenia, movement disabilities in Parkinson's disease—they always target one distinct part of the brain before spreading to affect larger swaths. My work has been guided by this principle, a fundamental tenet of "anatomical biology." Anatomical biology, a concept that was first developed in the nineteenth century and then validated in the twentieth, assumes that because each brain region houses a distinct population of neurons, individual brain regions are selectively vulnerable to different diseases. By using sophisticated neuroimaging tools to map the earliest stages of disease, my lab has applied the logic of anatomical biology to distinguish Alzheimer's disease from cognitive aging, isolate different molecular defects, and initiate therapeutic programs. Selective regional vulnerability has also been documented in a raft of other complex disorders—neurological diseases such as Parkinson's, Huntington's, and Lou Gehrig's disease, and psychiatric disorders like schizophrenia and depression—all of which initially attack one region of the brain.

So early on, when discussing autism with Dan, I insisted that no matter how complex its ultimate behavioral manifestations, it too should abide by the anatomical biology principle: there must be one brain region that acts as autism's anatomical source, its ground zero. Dan disagreed. Now I will admit that after decades of rigorous research, in which neuroimaging has been applied to many people with autism across their life spans, my position seems very unlikely. It appears Dan was right all along. While autism does not affect the whole brain—no condition ever does—it seems to defy the idea that there is just one selectively vulnerable brain region.

Dan's groundbreaking genetic work has, however, ended up showing a different kind of selective vulnerability: not vulnerability across the brain, but vulnerability within a neuron. Dan's research and the research from many other labs have shown that the proteins expressed by nearly all the genes in the network associated with autism function in a selective region within neurons: the dendritic spine. The premise of regional vulnerability is that once you identify where, you can then ask "Why there?" The promise of regional vulnerability is that if you can answer that question, you can potentially understand the mechanisms that underlie any brain condition. So, what are these altered proteins doing in the dendritic spine? It turns out that one central tendency of the network of autism-associated genes is to conspire against the molecular pathway that upregulates forgetting. On average and as a group, in people with autism the control knob for forgetting appears to have been turned down.

A reduction in forgetting can explain why some people with autism have exceptional rote memory, sometimes called savantism. "Savant" is a term used for someone who has one exceptional cognitive skill, in this case rote memory, like the Dustin Hoffman character in *Rain Man*. Among the few neuroimaging studies that have compared autism with and without savantism, most suggest that savantism is associated with a larger cortex and that within the cortex the thickest region is in the vicinity of the cortical central hubs. The kind of memory observed in savants is not the kind of memory seen in those with superior hippocampal function. Remember that the hippocampus functions in binding multiple components of complex events, distributed across multiple cortical hubs, to form new conscious memories. Close your eyes and think about your childhood bedroom. Thanks to your hippocampi, your mind can now pop up the three-dimensional space of the room so that your mind's eye can rotate in multiple dimensions: maybe clockwise to see your desk, reminding you of detested homework; maybe counter-clockwise to see your bedspread's pattern. You might notice that old lighting fixture as you look up, or a familiar patch of carpet discoloration (and the story behind it) as you look down. This recollective roaming of a cognitive space is what most typifies a conscious memory, which the hippocampus constructs by binding places with objects and time. But, again, this type of memory is not what is exceptional in autism. In fact, if anything, people with autism typically perform less well on formal hippocampal-dependent memory tests.

Exceptional rote memory is very different. Close your eyes again, but this time generate a list of words that start with the letter *t* and contain the vowel *o*. Let's assume that one of the words you come up with is "tool." No roaming of a recollected cognitive space, say a garage, was needed to generate the word, nor were any associations, say visualizing a hammer or thinking of an acquaintance who happens to be a carpenter. In other words, there was no need for associative recall, no reactivation of multiple cortical central hubs. With help from your cortical librarian, the prefrontal area, the word just automatically came to you, machine-like, as if you were reciting a word from a laundry list. Admittedly, nouns like "tool" are cognitively stickier than conjunctions like "though," as they easily associate with other features and events in time and place, and they might have snuck onto your list with some hippocampal help. In fact, a common trick of mnemonists is to exploit the hippocampus by embellishing a memory when asked to memorize by rote. These memory magicians rapidly learn dozens of items in one sitting by creating a fictional cognitive space around each item—a cognitive theater, connecting an item to an imaginary place staged with as many items as possible. But this hippocampal trickery is not how most of us recite dates, facts, and words, and it is certainly not how savantism works in autism. Rote memory depends on how well cortical central hubs function and, unlike associative memory, may depend very little on the hippocampus.

While these islands of enhanced cognitive skills might be fascinating and can confer some benefits, they are not found in the majority of people with autism. What *is* found in all people with autism is behavioral inflexibility, so much so that "repetitive and restrictive behaviors" is a required clinical feature when the diagnosis of autism is considered. Reminiscent of Funes never leaving his room, one of the reasons people with autism have behavioral inflexibility—why Freddy insisted on taking the same route home, or why Richard insisted on the same bedtime sequence—is plausibly linked to their diminished forgetting, their blunted memory chisel, and their difficulty sculpting preexisting memories stored in the cortex.

Animal studies support this interpretation. When genes are manipulated to express many of the genetic alterations that have been linked to autism—genes that are part of the forgetting toolbox—they cause dendritic spine growth and diminish forgetting. The impaired forgetting in these animals is associated with a strong preference for taking the same route out of a maze, over and over again—like Freddy and his preference for taking the same route to school—even if they would benefit from taking an alternate route.

Even if deficits in normal forgetting can partly explain the behavioral inflexibility observed in autism, they do not completely explain the obsession with sameness. If your forgetting, your memory chisel, has been blunted, you might learn a new route home more slowly and be frustrated by having to do so, and therefore prefer no change. But most of us would ultimately adapt, certainly if a behavioral change was beneficial. Even during emotionally charged childhood, most people do not insist on sameness with the vehemence exhibited by some children with autism.

Something else must be at play to explain why, as Kanner put it, children with autism seek solace in sameness, why small changes provoke such anxiety. In fact, there is another, more important benefit of forgetting to our cognitive abilities, one that better explains this need for sameness. Less obvious than simply helping sculpt our memories and increasing our behavioral flexibility, this cognitive ability is so deeply embedded that it is best appreciated only in its absence. For Borges's fictional character, the most paralyzing consequence of losing his forgetting was losing his ability to cognitively generalize. Whether it was a dog or any other object he was repeatedly exposed to—even seeing himself in the mirror—his mind registered each percept as novel and categorically distinct. It is true that with changes in lighting throughout the day, different items project different information onto the visual cortex. Nevertheless, most people's minds easily compute "same dog," "same person." Without the ability to forget, Funes lost the ability to generalize, to extract a gist or a gestalt—one of our most powerful cognitive skills—which helps explain an autistic person's need for sameness.

While experiments in mice and flies have most clearly

demonstrated the importance of forgetting for behavioral flexibility, computational science has best validated forgetting's role in our ability to generalize. Sift through your hundreds of digital photographs and find three of the same face. Your mind immediately recognizes the person despite the inconsistent visual information in each photo. Different lighting changes the color of the face; different angles shift its shape; different haircuts, headwear, eyewear, with or without makeup, all change the face's features. Despite this, your brain computes and recognizes "same person." Information processing in artificial intelligence (AI) was transformed when the design of computer algorithms started to borrow heavily from the workings of the brain. Facial recognition has emerged as a major field within AI because it is useful not only for Google searches or finding that one person buried in your photo library, but also for law enforcement. When computer algorithms began imitating how our cortex organizes the flow, processing, and storage of information, facial recognition in AI dramatically improved.

Facial recognition is an example of "recognition memory," which is procedurally different from "recall memory" in that you are shown a previously learned item and simply asked whether you recognize it or not. The synaptic plasticity that allows us to recognize a face occurs within our visual central hub. You do not necessarily need your hippocampi to recognize a face, because you do not need to bind multiple central hubs into a multicomponent conscious memory. When patients who have lesions that localize to the hippocampus, like H.M., are shown a previously seen face, they will consciously deny ever having seen that face before. But when forced to guess, they tend to guess right. The synaptic plasticity in the hub-and-spoke processing of the visual cortex, from lower hubs up to the central hub, are normal. Despite no conscious recollecting—no hub binding—the patients' conscious-less recognition is evidence that they can retain information stored within the hub.

The most successful computer algorithms for facial recognition are modeled on the hub-and-spoke processing that occurs in the visual cortex. In these algorithms, the face is first broken down into its component parts, and each part is encoded in a lower-level hub. The lower hubs, equivalent to the lower hubs in our visual cortex, code for primary colors and shapes, which then converge onto higher hubs, reconstructing individual facial features, and so on until the highest central hub in the facial processing stream "sees" the fully reconstructed face. Like neurons in our brains, each layer in the computer algorithm is made up of a network of spokes that connect together to code for an individual facial feature. In fact, computer science is so heavily influenced by neuroscience that each individual node—within a matrix of nodes that comprise a layer—is formally called a neuron. These artificial neurons abide by the same rules of synaptic plasticity as our natural ones.

The ability to recognize a person from their mugshot is nearly trivial, for us and for AI. First introduced with the advent of photography in the nineteenth century, mugshots were developed with an implicit appreciation of the computational challenges of facial recognition. Taken using fixed angles, lighting, and background, mugshots were designed to be easy on the eyes and on the mind. Not surprisingly, it is easy for a computer algorithm to recognize a face sifting through a database of mugshots. But, of course, we ask more of computers, as we ask more of ourselves. It has become a hackneyed trope of espionage movies to feature a supercomputer capable of zooming in on a face in a crowd and pinpointing a "person of interest"—in any kind of light, whether the face is contorted with fear or in a rictus of evil laughter, even if furnished with a wig or false mustache. AI's facial recognition ability is indeed improving, but our minds remain much better at identifying faces, which is why computers have not (yet) replaced humans in border control or airport security.

To understand this remarkable human capability, consider how one lower level in our visual processing stream codes one facial feature—say, a mouth. When presented with photos of the same person, the neurons in that level would have to recognize thousands of variations: smiling or scowling, cigarette dangling from the left or right, lipstick or not. The level would need to have sufficient computational flexibility to accommodate the multitude of possibilities confronted during sensory processing. As with behavioral flexibility, high memory capacity could theoretically help build the kind of computational flexibility needed for sensory processing—but only in a world in which the millions of variations are predetermined. One of the reasons why IBM's supercomputer Deep Blue beat chess grand master Garry Kasparov was that Deep Blue's vast memory capacity was able to store all plausible chess moves. The plausible moves for winning are finite and thus could be stored in the supercomputer's memory. But when it comes to pattern recognition, our brains can do better. Let's assume that we could endow the "mouth hub" in our brains with enough dendritic spines to match the storage capacity of a supercomputer, allowing it to store, say, every known lipstick color. Because our sensory processing streams are flexible enough to accommodate a potentially infinite number of small variations, the hub would still be able to recognize the mouth when painted with a brand-new lipstick color developed in the future.

Computer science has taught us how we do this. By testing different computer algorithms, computer scientists have learned that adding more memory—the equivalent of adding more dendritic spines—will not improve pattern recognition of faces or of anything else. Instead, the more effective way to artificially create human computational flexibility is to force the algorithm to have more forgetting. In computer science, this type of forgetting is sometimes called dropout, meaning a particular level is forced to reduce the number of artificial synapses dedicated to processing a facial feature—the digital equivalent of our own normal cortical forgetting.

To grasp how forgetting helps, zoom in on the mouth of a person in a photograph taken with a very high-resolution camera. Note the level of detail you can see and then consciously register every wrinkle in the lower lip, every unshaved hair above the upper lip. If you had enough dendritic spines in the mouth hub in your brain, you could potentially store all the information about this single photograph with pointillist precision. If you had this level of photographic memory of the mouth, after one viewing you could recall and re-create—if you had the artistic ability—the mouth with high fidelity, a rote memory feat performed by some people with autism. Computer science has taught us that while this would be a remarkable achievement, it would come at a high cost to your computational flexibility and ability to generalize. Obsessed with the smallest details, your mind would be unable to recognize the same mouth even with subtle variations. You would get stuck at this lower-level hub, which would stem the flow of information processing to higher hubs and slow the reconstruction and recognition of the whole face.

Computer scientists have learned that they can overcome this problem by preventing such a high degree of photographic memory. By building in the active forgetting we use at every level in the computer processing stream, the engineers ensure that the computer's layers record and store just the gist, not every detail, of a person's facial features. Forgetting is required so that each hub can store just enough information to recognize, but also to generalize, each facial feature and ultimately the face as a whole.

With autism spectrum disorder comprising a heterogeneous collection of conditions, and with studies investigating people at different times during this evolving condition, complete consensus in autism behavioral research is rarely achieved. Nevertheless, the vast majority of studies confirm that sensory processing in autism is characterized by a predilection toward lower hubs, toward seeing the trees instead of the forest, consistent with what is observed in computer algorithms that don't incorporate forgetting in the lower layers of the processing stream. These psychological studies have validated Kanner's clinical hunch that in autism there is a "persistent preoccupation with parts of objects."

One of the most elegant studies was inspired by the paintings of Giuseppe Arcimboldo, the sixteenth-century Italian artist who created portraits composed of fruits, vegetables, and flowers. The illusion of seeing a face in this visual salad exploits the way our visual processing streams work in sensory integration—that is, our tendency to synthesize parts into a whole, which is so strong that it often fools us into seeing faces in clouds, rock formations, or even the grille of a car.

The investigators created a series of stimuli comprising different combinations of fruits and vegetables on a plate. But unlike Arcimboldo's portraits, in which one can't help but see a face, the stimuli varied in how face-like they were. They administered the stimuli to groups of children with and without autism. Those with autism, on average, took longer to recognize faces in the stimuli. This delay was interpreted to have occurred because the autistic children fixated on each item on the plate, slowing their minds' ability to integrate the parts into a whole.

You can try playing with food on a plate to re-create these stimuli and get a sense of the study. Place a small strawberry in the middle of a round white plate where a nose might be. Add two carrot slices above and on either side of the strawberry for eyes, then a melon wedge right below it for a mouth. Add two apple peels above the carrots as eyebrows. Take a picture and share it with others. If done right, this "recipe" will cook up a face in the visual cortex of nearly everyone looking at it. Now jumble the items or remove some of them, varying how face-like the arrangements appear, and take pictures of them. Choose one picture that, though difficult, most people would ultimately recognize as a face—perhaps one where the strawberry is removed or one apple peel is transposed with a carrot slice or the melon wedge. Show this to different friends and record how long it takes each to recognize the stimulus as a face. The one who recognizes it as a face the fastest presumably has the least sticky dendritic spines in the lower-level visual cortex hubs. In the slowest one, these dendritic spines act like Velcro, slowing their ability to integrate the whole.

Another, older study that used real-life stimuli measured the time it took for subjects to complete a jigsaw puzzle. To get a sense of this study, imagine if I spilled out hundreds of jigsaw puzzle pieces on a table. In one case, I let you keep the box, which has the picture of the finished puzzle on it, and you are allowed to use this "whole" as a guide in putting together the parts. In another case, I don't let you keep the box. Obviously, you would benefit from seeing the finished product and complete the puzzle faster in the first case. This study has shown that, on average, people with autism benefit less from seeing the box than people without autism. In fact, some autistic subjects completed the puzzle in the same time with or without the box—piece by piece, part by part, seemingly oblivious to the whole. For some people with autism, even when shown the forest, they remain preoccupied with the trees.

Some psychologists have extended this autistic bias toward the parts at the expense of the whole to help explain another clinical feature required for making a diagnosis of autism: "persistent deficits in social interaction and social communication." Socializing also depends on pattern recognition, but now the pieces are the constellation of social cues that the person with whom you are interacting is continuously projecting. Instead of synthesizing facial features to recognize a face, you need to synthesize a range of cues to recognize the person's social intent. Is that smile genuinely friendly or merely polite? Is that tone earnest or sarcastic? Subtle differences in how the algorithms in your mind first deconstruct these complex social cues and then reconstruct them into a global interpretation will influence your response. This back-and-forth is the essence of socialization, and how well you engage in this social discourse will determine whether you are considered socially savvy or awkward. Processing incoming social stimuli, while harder to map anatomically, likely follows the same hub-and-spoke pattern as processing facial features. The local processing bias that exists during the sensory processing flow might, therefore, also explain the difficulties people with autism tend to have with social interactions.

Due to the convergence of insights from computer science and the autistic condition, we now understand that forgetting allows us to better record and recognize representations of the outside world. Both artificial and our own natural intelligence depend on forgetting in order to generalize—to reconstruct the whole from its component parts so that we can categorize and label things even when they vary subtly in infinite ways.

Philosophers can debate how well our minds faithfully reconstruct and mirror the outside world; magicians will continue to exploit this innate ability by leading us to falsely reconstruct patterns from what we simultaneously see and hear. Regardless, most of us want to be able to knowingly conclude that a dog seen in the morning and in the evening is the same dog. Surprises might be nice on occasion, but imagine being continuously surprised by everything you see or hear. At some point, the endless shock and awe would cause psychic discomfort. Think back to a large, bustling event where you might have experienced too much unrelenting novelty. For me it would be going to Times Square one New Year's Eve. While I at first enjoyed the cacophony, the bright blinking lights, the chaos and novelty of it all, it ultimately became discomforting and even anxiety provoking. What a relief it was to return to the familiar quietude of my small apartment. Weighing unabating stimulation against tranquil familiarity, one can understand why a mind with diminished cortical forgetting might much prefer sameness. The capacity for generalization that comes with cortical forgetting allows us to better organize and catalog, and thus arrange the clutter and squelch the clang of an external world sensed only as parts.

As succinctly put by Jorge Luis Borges: "To think is to forget a difference, to generalize, to abstract." Autism has shown us how challenging life can become if the balance between memory and forgetting is thrown off-kilter by reduced forgetting. As clinically put by Kanner: "Autistic children show a peculiar type of obsessiveness that forces them to postulate imperiously a static, unchanged environment.

Any modification meets with perplexity and major discomfort. The patients find security in sameness, a security that is very tenuous because changes do occur constantly and the children are therefore threatened perpetually and try tensely to ward off this threat to their security."

A mind without forgetting might thrive in a world without flux. But we now know that minds that forget in balance with remembering have ideally evolved for our fluidly, sometimes turbulently, fluctuating worlds. Thankfully, all of us across the spectrum—all of us—have some degree of forgetting. Because a mind with no forgetting would be paralyzed by an unbearable desperation to flatten and fixate the world into sameness.

THREE

LIBERATED MINDS

Dr. Yuval Neria, a professor in Columbia University's Department of Psychiatry, directs the program on post-traumatic stress disorder (PTSD). One of Israel's most highly decorated soldiers, he received the rare Medal of Valor, the military's highest decoration, for his extreme acts of bravery as a tank battalion commander during the 1973 Yom Kippur War. I first met Yuval after he joined the faculty in 2011, but I'd known about him since my youth. I grew up in Israel, emigrating from the United States with my family in 1970, and at the time war heroes were the country's celebrities.

Shortly after his recruitment to Columbia, Yuval reached out to me to explore a potential collaboration on the link between PTSD and memory. With his military reputation, I wasn't sure what to expect, but when we met I discovered that Yuval is what we sometimes describe as a "golden" Israeli. "Golden" in this context does not mean the glitzy, brashly arrogant stereotype some people have of Israelis, but in fact just the opposite. When I was growing up in Israel, "golden" referred to an exceptional person with a deep sense of humanism and compassion, a personality blend of humility and quiet strength. In retrospect, I should not have been surprised that Yuval had these qualities. In mentioning him to any of my Israeli friends, they reminded me that after his famous military service, he became a founding member of the grassroots Peace Now movement, whose main mandate was to reconcile the decades-long conflict between Israelis and Palestinians. He also wrote a novel, inspired by his wartime experiences, whose astute psychological insight rests on his acquired wisdom about the suffering caused by life's inflicted traumas, big and small.

Yuval knew that my lab had developed the right sort of MRI tools to investigate the anatomy of memory, but he had no idea that I had grown up in Israel. Once I told him, our scientific discussions about how memory and painful emotions connect in the brain often digressed toward the personal and toward Israel, or "the land," as it is sometimes referred to in Hebrew. A common couplet of questions posed by native-born Israelis to immigrants like me, as a way to establish our Israeli bona fides, is whether we served in the Israeli military, and if so, where. Yuval asked, and I answered: "yes," and "in Sayeret Golani," one of the military's special operations units. In most cases, the conversation about my military days would happily go no further. But Yuval was very familiar with one of the unit's most famous operations, the battle for Beaufort Castle, and he wondered whether I had participated in it. I had.

A fortress in southern Lebanon built during the Crusades, the Beaufort castle sits high atop a mountain cliff overlooking Israel's northern border. During the late 1970s, farmers, schoolchildren, and other civilians in Israel's northern Galilee region had become frequent targets of rockets launched from the Beaufort, and for them—and Israel at large—it was no longer the "beautiful fortress" that its French name implied. On the morning of June 6, 1982, Israel's defense forces initiated the First Lebanon War. Our unit had been sent as the vanguard into Lebanon the night before to secure the Beaufort, which meant taking over a series of trenches surrounding the castle that were manned by Syrian commandos. The trenches had been designed with deadly details—narrow passages with extraordinarily high concrete walls, a complex labyrinth with internal bunkers, and external fortified positions outfitted with machine guns and rocket-propelled grenades. Trench warfare, with its close-range gunfire and hurtling explosives, is typically one of the goriest types of battle, and that night at Beaufort Castle was no different—so gruesome, in fact, that I refuse, as I have done ever since, to go into its bloody details.

Much about the battle is public knowledge, but Yuval seemed to know a lot more about it, and I suspected that he might have gained some inside information from the many military generals he was still in touch with. At some point—inevitably, given his own experience and his clinical training, and since we were joining academic forces to study memory in PTSD—Yuval asked me if I or any of my military friends had ever suffered from it. It's hard to believe, but despite the growing awareness of PTSD and my own medical training, my comrades and I had never broached the topic. Whenever we got together, we invariably returned to those painful memories, but only cloaked in the guise of gallows humor.

Although we'd sworn no oath of secrecy, there was a general understanding that not just the details of the battle but also what happened to us a few months later were best kept among ourselves. At Yuval's prompting, I reached out to some military friends and was given permission to share some of our post-battle memories. The war raged on, but after completing a few more special operations, our unit was ordered to return to our home base in northern Israel to await further assignments. Secure and secluded from the rest of the military, we lived together in this pressure-cooker limbo, awaiting our next assignment. Our rectangular concrete barracks, surrounded by large eucalyptus trees, were a holdover from the British era. The base had been retrofitted by Israel's military to become an elite school in the arts of reconnaissance, special weaponry, self-defense, and killing. Thankfully, no more assignments materialized, and we were discharged a few months later when we reached the end of our tour of duty.

Many of us exhibited altered behaviors during those last few months. Before the war, none of us were serious alcohol drinkers or recreational drug users (at the time, neither alcohol nor drugs were popular among Israeli teenagers). Now we indulged for the first time in whiskey and vodka, bottles of which we kept secretly stashed in our military-issue gray metal dressers. Some in the group began experimenting with marijuana. A small number of us, who loved jazz and literature, started blaring Coltrane while writing and enacting what we thought were theater of the absurd skits. The unit commanders thought we were just acting out, and so did we. However, when one of our skits involved lewd behavior while wrapped in the Israeli flag, the leadership became concerned. I vaguely recall a discussion about bringing in a military psychiatrist, but nothing came of it. PTSD was not yet a well-known disorder, and we all chalked up our actions to stress.

In sharing these memories with Yuval, and perhaps as a subconscious ploy to hide my embarrassment for never addressing the obvious clinical question, I expressed some ignorance about PTSD's formal criteria. Yuval knowingly smiled and professorially reviewed the criteria. Shortly thereafter, I got in touch with two of my closest friends from the time, and we reviewed these criteria together. We went down the list with eerie detachment, wondering whether any applied to us, as if we were filling out a customer service survey. At a subsequent meeting, I discussed my "findings" with Yuval.

Symptoms of PTSD, which typically start a few months after the traumatic event, are grouped into four general categories. One is avoidance of the traumatic event, which was true for us, with the exception of our occasional get-togethers. Yuval told me that this behavior is often the case in PTSD triggered by wartime trauma, a typical dynamic among comrades-in-arms. Another category of symptoms is chronic negativity about oneself and the world and a gloomy hopelessness about the future. All three of us, it turns out, are considered cynics by our families (we don't agree!), but none of us is bleak and hopeless. A third category is heightened emotional reactivity—for example, startling easily and hypervigilance regarding perceived danger, an emotional state that can lead to trouble sleeping and angry outbursts. We all dislike fireworks displays because they provoke discomfort, and we all immediately and uncontrollably identify emergency exits whenever we enter a confined public space such as a theater or stadium. These thoughts are run-of-the-mill, however, and do not seem pathological.

The final category, the one with the greatest relevance to this book, is a defect in what is called "extinction," a psychological term that describes the ability to forget the trauma. As Yuval explained, this category is the most critical for diagnosis and drives most of the other symptoms. It is characterized by "intrusive," or recurrent and distressing, memories of the traumatic event—flashbacks, nightmares, and severe emotional distress when exposed to something reminiscent of it. While we all have vivid and painful memories of the battle that still haunt our dreams, these memories do not cause us emotional distress and do not seem to meet the threshold required for extinction.

The ultimate diagnosis of PTSD depends on whether and how some or all of these symptoms clinically manifest—that is, if they are clearly a detriment to one's life. Even in our most self-critical states, none of us comes close to endorsing this clinical criterion. We are all happily married and have what we consider successful careers and family lives. So while we three were exposed to a clearly traumatic event and have indelible memories of its gory details, none of us, in Yuval's estimate, have or have had PTSD. Why not? To answer this question, we need to look at the brain mechanics of emotional memory—how emotions, particularly negatively charged ones, become part of our memory networks.

We've already seen how the brain forms new memories and how different sensory elements are bound together in becoming a memory; how each element is processed in different cortical streams, ultimately encoded in distinct central hubs; and how central hubs in the cortex interconnect with the hippocampus, which then binds them during memory formation so that they become integrated into a memory network. The next time you unexpectedly run into someone you know, pay attention to the mechanics of remembrance. Upon seeing the person, you can almost sense the pebble-splash of recognition and its outward rippling as the network of associated hubs is reactivated. The person's name emerges, but so does the wider web of related sensory details. Notice how the memory is often colored with emotions, which tend to be particularly vibrant and immediate when your previous experience with the person was negative. In some cases, the emotional component can be neon bright, so strongly connected with the memory that it is re-experienced even before many of its sensory details are revivified and come into full focus.

Emotional memory, particularly when negative, serves a clear advantage in helping us adapt to our worlds. Yes, our complex worlds are "blooming and buzzing," a term introduced by the father of American psychology, William James, to describe the confusion young children must experience when their new minds start processing the profusion of incoming sensory information. But blooming worlds are thorny; buzzing worlds sting. Remembering whether a person is friend or foe, or whether a situation should provoke fear for flight, is required for our survival. In worlds less physically violent, where mortality is not immediately at stake, the emotional component of memory is still useful for social survival, as any middle schooler can attest.

Danger detection, therefore, is fundamental for living, and all living things have built-in, highly sensitive danger detectors connected to elaborate internal security systems. Mammalian brains have evolved an ingenious security system fueled by the "hypothalamic-pituitary-adrenal axis." When danger detectors go off, the hypothalamus, a structure deep in the brain stem, stimulates the pituitary gland to release chemicals into the bloodstream that coax the adrenal glands to release hormones, including cortisol and adrenaline. These "stress" hormones place our bodies on red alert, signaling systems to prepare for battle, retreat, or a strategic posture in between. The brain region most receptive to these stress hormones is the amygdala. Just like the hippocampus, we have two of them, and these almond-shaped structures reside just below the cortex (they are "subcortical"). Effectively the nervous system's "central command" for handling perceived threats, the amygdala uses its widespread pattern of connections across the brain to integrate relevant information and oversee, orchestrate, and mobilize the many divisions that make up our security systems, including connecting back to the hypothalamic-pituitary-adrenal axis. By establishing this critical closed loop, the amygdala can, when needed, ratchet up and amplify the alert signals, blaring out "Code red! Code red!" and triggering a panicked scramble.

If factual information is processed and coded by the central hubs in the cortex, the amygdala can be considered a subcortical central hub that processes and codes emotional information. Just like the cortical hubs, this subcortical one connects with the hippocampus, our memory teacher. Subcortical, emotional information, therefore, is incorporated into a newly formed memory alongside factual information from the cortex. In this way, the amygdala paints the bland facts of our memories—items, time, and place—with emotional color. And it turns out that the amygdala's emotional palette is most vividly experienced when spray-painting our memories with unhappiness, fear, rage, or misery. The adage "Happiness writes white"—conveying the idea that pleasant happiness, unlike melodrama, does not display vibrantly on the page—is as true in our brains as it is in fiction.

A few months after we conquered Beaufort Castle, our generals decided to arrange a tour of the battle site for the families of our dead comrades. I suppose this was meant as an act of commemoration. In retrospect, however, a family tour, which included younger siblings of the fallen and required crossing a hot border into enemy territory to a battle site still sullied with blood, seems to have been misguided. Confounding decisions of this sort are not atypical, I suspect, for war-worn countries in which combat is embedded in their

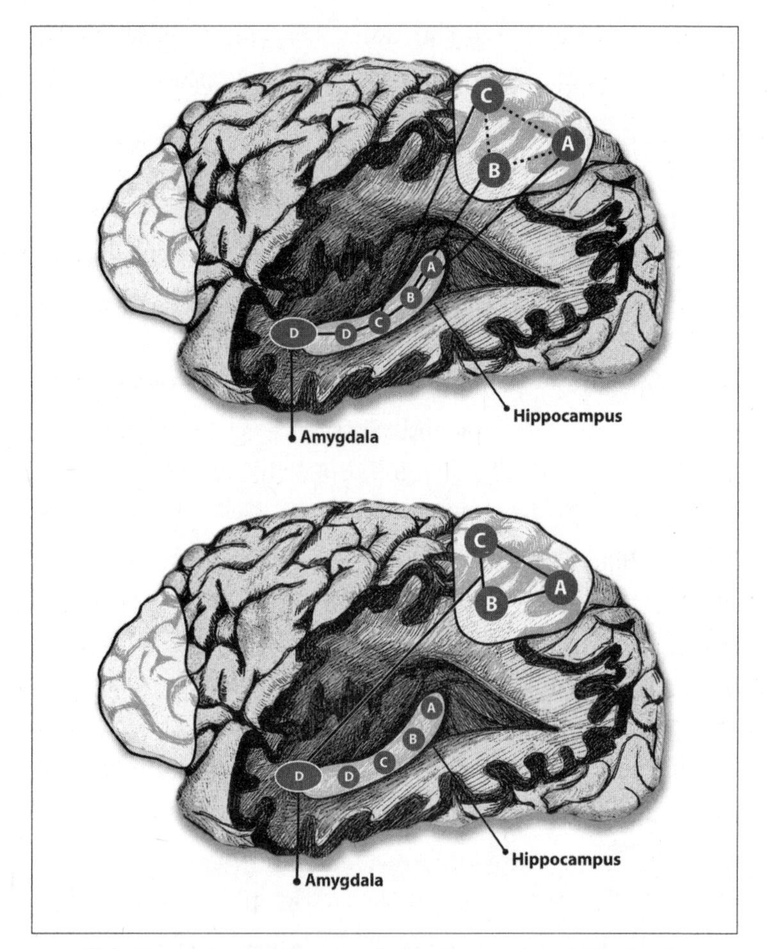

The amygdala and the emotions of memory: at the top is during hippocampal training, and below it is after.

ethos and every generation has experienced war or can expect to. We peacefully strolled the grounds of the battle site, the medieval castle now looming less ominously, once again the historical tourist attraction it was meant to be. The trenches, in the light of day and without the drumbeat of war, looked like dry irrigation canals. A warm easterly breeze came up from the Litani River, the surrounding shrubbery dusty after a dry summer. We politely answered the families' questions. They asked about the battle in a strangely formal way, as if not asking might dishonor the memory of those lost or somehow be misinterpreted as a lack of interest. We answered carefully but not completely. By implicit consensus, we had decided that some details were best left unspoken.

Here we were, just a few months later, in the very place where our hippocampi and amygdalae seared our minds with red-hot memories. Undoubtedly, the full memory network encoded during that cursed night was reactivated. Yet despite some unease, a gnawing in our stomachs, we were not disabled by the memory, nor did we manifest any symptoms of high anxiety. Normal forgetting had, apparently, already kicked in.

Try flipping through a yearbook and finding a picture of the schoolyard bully. Your emotional response, while still negative, has probably been dampened by the passage of time: the normal forgetting mechanisms described in chapter 1 have had their beneficial effects. If this normal forgetting is defective, psychopathology can occur—phobias and other anxiety disorders such as PTSD. In these disorders, the full force of the memory network is reactivated, provoking a heightened and incapacitating emotional response. Let's assume that the photo of the bully triggers in you, even decades later, the same—or at least a similar—emotional response as back in the schoolyard and you reexperience the full force of your fear. Or perhaps the photo reactivates the violence you felt on that one day you fought back, your ability to think clearly, now as then, clouded by bloody-minded rage.

Clinicians like me—brain mechanics of a sort—can now ask which part of the brain is faulty when this occurs. It is possible that the whole memory network might be hyperactive—the cortical hubs that stored the bully's face, his name, the time, and the place, together with the subcortical hub that imbued this memory with the neon colors of either fight or flight. Or, possibly, a select few of these hubs could have become overly connected all these years later. We might even consider the hippocampus to be the anatomical source of the psychopathology. Although normally the hippocampus is not needed to recall a decades-old memory, it is conceivable that in this psychopathology of "too much memory," the hippocampus could remain abnormally and chronically hyperactive, throwing the memory network into overdrive.

Recent functional imaging studies have shown that despite the distinct details of each phobia or traumatic event, the anatomical source of impaired emotional forgetting is typically the subcortical hub: a chronically hyperactive and hyperresponsive amygdala. I've heard that some of my comrades with whom I've lost touch did suffer from symptoms that might be considered PTSD. This raises the interesting question of why, when exposed to an identical traumatic event, some people develop PTSD and others do not. Many factors predispose some people versus others to develop diseases when exposed to similar environmental hazards—consider smoking and heart disease, for example. The same is true for PTSD. At a cellular level, the pathology of PTSD might be reduced to malfunctioning amygdala neurons, which appear chronically hypersensitive and hyperresponsive. Similar to the mechanisms described in chapter 1, the amygdala's neurons are capable of synaptic plasticity and dendritic spine growth. The larger and denser their spines, the more responsive the neurons are to incoming stimuli. One idea that can explain why some amygdala neurons undergo a pathological growth of spines and others do not is that they are recurrently stimulated by red alert signals at abnormally high rates. This leads to a ratcheting up of spine growth until some threshold is reached, at which point those neurons shift permanently into a pathological state of chronic hypersensitivity, which in neurology is sometimes called a spastic response.

The general approach in treating PTSD is to exploit the mechanisms of normal forgetting to reprogram the amygdala and shift its activity back to normal. This is the logic of exposure therapy, where patients are exposed over and over again to an anxiety-provoking stimulus in a benign setting and where normal forgetting mechanisms are activated, overriding the chronic state of hypersensitivity. If simple exposure is insufficient, in-depth psychotherapy can disentangle a traumatic memory that might have become ensnared with other highly charged emotional memories that continue stimulating the amygdala. Cognitive behavioral therapy can also be effective by helping a patient become aware of the faulty thought patterns that often misinterpret or exaggerate emotional cues. All these interventions are designed to defang the memory, to silence its alarms, to relax the spastic neurons—although I suspect that practitioners of these sophisticated psychological interventions might object to this reductive cellular interpretation. If necessary, such behavioral therapies are often coupled with drugs that are known to dampen amygdala activity, which can further revitalize normal forgetting.

When Yuval and I discussed why some of my comrades and I did not suffer from PTSD, he was particularly interested in those intense few months after the battle during which we were huddled together in our unit's home base, "acting out" as we counted down the days to our military discharge. Was there anything during that time that might have inoculated us against PTSD? Possibly, according to Yuval, who noted how our teetotaler brains had been suddenly exposed to alcohol, which is known to dampen amygdala activity. For obvious reasons, high alcohol consumption is not a clinical recommendation, but for some of us, during that particularly fragile time, freshly back from battle, it might have been beneficial. Currently, researchers are testing other drugs that when taken in controlled clinical environments might help control PTSD, including MDMA (ecstasy) and even LSD.

None of us used those drugs back then, but some started using marijuana. Cannabis contains an interesting group of chemicals, primarily tetrahydrocannabinol (THC) and cannabidiol (CBD). The brain contains specific receptors for THC, and consumption of it stimulates the amygdala. The dread and anxiety sometimes associated with pot can be traced to the amygdala, which is very likely studded with higher concentrations of these receptors. While there are no specific receptors for CBD, it binds other receptors that are known to dampen amygdala activity. So it's possible that for some of my comrades, perhaps those less sensitive to THC and more sensitive to CBD, pot smoking was beneficial.

Yuval was also interested in our absurdist skits. I told him how some involved morbid themes using a range of props. One prop was procured during what we farcically called a "nighttime raid" on our own air force's helicopter landing strip, conveniently located just outside our base. When most of the base was asleep, we snuck up to the landing strip and seized an American flag from a pole hastily planted in anticipation of an impromptu visit by Caspar Weinberger, the U.S. secretary of defense at the time. The following day, we used both the American and Israeli flags to enact a binational funeral procession. The skit was meant to satirize what we perceived to be the Reagan administration's over-coziness with Israel, enabling what we considered a misguided war. As I listened to my own voice recounting the details, it struck me how the skit now seemed more silly than satirical. But for Yuval what mattered was that we had engaged humor, whether sophisticated or sophomoric. He explained how the skits probably functioned like exposure therapy: as we played out emotional elements of our memories over and over again, we bathed them with humor, bleaching out their bloody hue.

Most important in Yuval's view, we lived together in that fraternally intense and highly communal environment for months after the battle—comrades-in-arms, sometimes literally. Apparently one of the greatest risk factors for PTSD in soldiers is finding themselves alone and lonely shortly after the trauma, their minds exposed, without a protective social fabric, to the lashing loops of their misery, dread, and fear. Here too there might be an interesting neurobiological correlate linked to the amygdala. Interacting with loved ones—and there was no question we were in brotherly love—causes the body to secrete oxytocin. The amygdala, a brain structure uniquely receptive to all sorts of emotional signals, is highly enriched with oxytocin receptors. When oxytocin binds these receptors, the amygdala's activity is dampened, which is one reason oxytocin is known to mediate the intense social bonding with those we love.

Emotional forgetting doesn't simply reduce the risk for developing psychopathologies. It also frees us from the prisons of pain, anguish, and resentments—even the small ones that accumulate and fester in any interpersonal relationship. Marriage therapists have told me that even the happiest life partners could benefit occasionally from a pill that would help them forget emotionally, and in fact some therapists prescribed ecstasy to their patients for a time until it became a banned substance.

More broadly, normal emotional forgetting liberates us from the drag of those personality traits that all agree are ugly, counterproductive, and embittering—traits that can be categorized as the amygdala's deadly sins: spite, vindictiveness, malice, vengefulness, and even (the one I most detest) righteous indignation. I am quite sure that whenever any of us veer into that sinful terrain, the gears of our amygdalae are whirring and whining in overdrive. Lastly and most loftily, emotional forgetting, in freeing our minds, allows us to forgive. Forgiving does not, and should not, entail the factual forgetting of an offending event. But letting go of seething resentments is required for forgiveness—the noblest example of the benefits of a forgetful mind.

We should remember to try to emotionally forget, just enough, for our sanity, for our general well-being, for the sake of our family and friends. Easier said than done, I know. As a doctor, I can't formally recommend using recreational drugs that relax the amygdala and its built-in propensity for spasticity. I can, however, recommend talk therapy of all kinds, whether with a therapist or simply conversations among friends. And as a non-golden Israeli who can't shake some of his abrasive tendencies, as a neurologist who has been indoctrinated to treat pharmacologically, and as a neuroscientist who tries to reduce many things—even sometimes absurdly—to molecules, I now appreciate a simpler and more elegant way to enhance our innate capabilities to emotionally forget: socialize, engage life with humor, and always, always try to live a life glittered with the palliative glow of love.

POSTSCRIPT

In writing this chapter, I not only needed permission from my military friends to share some of my wartime memories, but I also needed their recollections for fact-checking. As a memory researcher, I know all too well that memories are continuously sculpted, and after long passages of time, the creative mind abstracts, contorts, and even distorts the past. Familiar with the pitfalls of nostalgia, I know that our minds store the past less like a museum of personal history than a gallery of memory art. So I double-checked my wartime memories with my friends. While doing so, I was informed by one of them that, "oh, by the way," if evidence was needed, he had that American flag—the one we pilfered from our helicopter landing strip. This friend was born and raised in one of Israel's religious kibbutzim but left the country a few years after his military discharge. Now a secularized Jew, he wandered the Diaspora doing different odd jobs, ultimately settled in New York to raise a family, and currently, to my delight, was living a few blocks away from me in Manhattan.

"Let me get this straight," I responded with astonishment. "You not only decided to take this flag home on some weekend leave while we were still in the military, but once you left the country, among the limited items you traveled with from place to place, this flag was one of them?"

"Yup," he bluntly replied, as if this was obviously a treasure worth dragging around with him throughout his itinerant life. I guess it was.

Another friend, who still lives in Israel, part of our trio of army brothers, was scheduled to visit the States a month or so later. He, like me, was equally stunned to learn that the flag was in safekeeping. One evening during his visit, the three of us met at my apartment. We removed the flag from the brown bag in which it was carefully folded and unfurled it on my kitchen table. The last time we had viewed this flag together was a lifetime ago. The scene in my apartment had all the makings of a dramatic ending to a documentary film, when an archival search digs up something startling, something that carries with it the promise to jolt.

The reality was that this particular memento was met with a flat fizzle. Perhaps it was because a flag is inherently generic, or perhaps it was a lesson in how souvenirs are often neutered of their associative capacities: not every element of a memory is sufficient kindling. We experienced that evening the common disappointment of a forced recollection, the kind of anticlimax commonly experienced when, say, attending a class reunion with anticipation or sifting through a recently discovered photo album in search of lost time. Sometimes memories are best left alone, viewed only in the gallery of our minds.

We did, however, realize something we'd somehow failed to notice so long ago. The flag must have been rapidly handsewn: the stitching was knotty with imperfections, and the stripes were so hastily fastened that the bottom of the flag revealed the white linen fabric onto which the last red stripe had been imperfectly aligned. Caspar Weinberger's visit had been impromptu, and it dawned on us that some poor soldiers had scrambled to sew this flag from scratch, likely the night before his arrival. While the flag provided proof that the event had actually happened, it now seemed more like unintentional folk art than a treasured memento.

Since then, whenever we talk about that wartime memory, that new encounter with the flag is one of the first things that comes up. If anything, that experience reshaped our memories even further, chiseling away at the pain of the original event.

FOUR

FEARLESS MINDS

We've already seen how deeply intertwined fear and memory are in the brain and how emotional forgetting is important for our psychological health. To further unpack our emotional memory, and to see how emotional forgetting is more generally beneficial, let's consider two cousins, C and B.

All agreed that C was very clever, but it was his ruthlessness that truly distinguished him. As a youth, C never backed down from a fight. Now, as a barrel-chested, gravel-voiced adult, he was always challenging authority—not with the nuanced repartee of a skilled diplomat, but with bloody-minded rage. Status obsessed and with no time to play, he rapidly ascended to the pinnacle of his society, one that rewarded his macho bullying. Unloving and unlovable, he nevertheless sired offspring from a collection of different mothers. Among his family he cultivated a reputation as a harsh disciplinarian, but his wrath was most easily provoked by those outside his social circle. He was a shameless xenophobe. They never met, but C undoubtedly would have sneered at his cousin B, whose friendly personality and social lifestyle C could never respect and would scarcely recognize. B had always been relaxed and tolerant, quick to forgive and happy to console. B empathized with friends and outsiders alike. As long as his community was harmonious, he seemed not to care less about social hierarchy or whether its leaders were male or female. While B's affability extended to the workplace, he dedicated nearly as much time to work as he did to play and romance.

Most of you will recognize these two types of diametrically opposed personalities in people, or at least fictional characters, you know. But C is not actually a mob boss, nor is B an enlightened humanitarian. C is a chimpanzee, and B a bonobo.

The zoological tree of life was updated during the 1970s, when taxonomists started relying on animals' DNA rather than external or internal appearances to classify them. Appearances do matter, and since they partly reflect the blueprint encoded in our genes, many of the older classification schemes turned out to be right. The greatest shock arising from the new taxonomy was regarding humans, and not only because we were the only ones paying attention. It stripped us of our anointment as sovereigns of the animal kingdom. We suddenly had to share our throne with our two closest living cousins, chimpanzees and bonobos, with whom we share more than 99 percent of our genes, and with whom we lived for millions of years before *Homo sapiens* diverged from the two *Pan* species, which subsequently diverged from each other. Our three species are so closely related that some have argued that we should all be reclassified under a shared genus, either *Pan* or *Homo*.

We might wonder, therefore, which of our cousins we are closer to. Our facial structure looks a bit more like that of the bonobos, as does our bipedal walking. Although we have a long way to go, our leadership—within families, clans, or society—can be matriarchal, like the bonobos'. But our social temperaments are much harder to measure, because personality is so heavily influenced by environment. For instance, bonobos raised in the wild are more aggressive than those raised in captivity, though never as ruthless as chimps. Compared with chimps raised in similar habitats, bonobos are naturally inclined to be more altruistic, compassionate, empathetic, kind, and tolerant. They retain childhood's playfulness into adulthood, and, borrowing from the primatologist Frans de Waal's many descriptions, they engage in more lovemaking than war. Collectively, social scientists call these friendly behaviors "prosocial" because they are thought to benefit society as a whole.

Be honest: Which of these extremes in personality describes you best? I have no doubt biased you in the way that I crafted the story lines of C and B. In truth, we should judge neither, since their behaviors have been appropriately shaped by the evolutionary pressures of their environments. But even if you are critical of C, it's fair to assume that many of you want to climb the rungs of your social, financial, or career ladders. Despite moral teachings you may have received, some of you might occasionally achieve your status on the backs of others. Even if your first inclination was to find B likable, I doubt that most of you are pacified to the point of never experiencing near-homicidal rage. The most ruthless among us still enjoy social activities and want to play and to love. In reality, most of our social temperaments, whether innate or more likely shaped by our early life experiences, represent some mixture of chimp and bonobo.

But let's assume that, to your dismay, you tilt more chimp than you would like. You have rage issues, sometimes act coldheartedly, or suffer dark bouts of misanthropy. Most troubling is your loneliness, your difficultly in connecting with others or loving unconditionally. How might you go about lightening up your temperament in the hope that it might improve your social life? The neuroscientist's answer would be to first understand the brain mechanisms that govern social temperament, a prerequisite for knowing the surest and safest ways to pull the levers that can bring about change.

One first step would be to identify brain areas that are linked to social temperament. Assuming you want to focus on normal variations in personality, as opposed to abnormal personality disorders, you might consider taking a closer look at chimps and bonobos, which can be considered evolutionary twins who were reared apart, in diametrically opposed environments that reward either antisocial or prosocial behavior. In 2012, researchers accomplished exactly this when they finally managed to complete an MRI study that compared large groups of chimp and bonobo brains, and the results were striking. Since it's easy to distinguish chimps from bonobos by looking at their external anatomy or by watching their many different behaviors, you might expect, as I did, that it would be easy to distinguish chimps from bonobos by looking at their internal neuroanatomy—after all, mammalian brains have hundreds of different regions and structures. This was not the case. Against my expectation, only a handful of regions reliably differed between chimps and bonobos, and all had been previously linked to social behavior.

One structure stood out the most: the amygdala. So clear were these MRI results, which were then confirmed by examining the brains of chimps and bonobos under the microscope, that the researchers concluded that the amygdala might be a key structure linked to social temperament. These studies, crucial for mapping a correlate of normal temperament, agree with imaging studies in human patients that have implicated the amygdala in antisocial personality disorders.

So we return to the amygdala, the brain's central command for recording and orchestrating our responses to danger from the outside world. The amygdala—that brain structure that learns how to perform these danger management functions through experience, by remembering or forgetting fearful memories. We have already seen that the amygdala's forgetting mechanisms can help us forget certain aspects of traumatic fear memories, thereby preventing antisocial behaviors that develop in some patients with PTSD, including raging and even occasionally violence. It's fair to wonder whether forgetting some everyday fear memories—not just the traumatic ones that cause psychopathology—might result in a friendlier disposition. But before we jump to that conclusion, we need to know a lot more. For one, we need to understand whether and how fear memories are related to rage and the rest of our misanthropic chimp-like traits. More important, to verify the hypothesis, as plausible as it might seem, we need to know how to induce forgetting of our run-of-the-mill fear memories and then show that this forgetting somehow improves our normal social temperaments.

To clarify all of this, we need to tell the amygdala's full story—about how it was discovered, how it functions, and how we have learned to induce fear forgetting.

As we saw in the previous chapter, it is owing to the amygdala's splashy eagerness to form negatively tainted memories that most of us can easily remember a childhood bully. Consider the responses he might have provoked. You might have stopped dead in your tracks at the mere sight of him. If he also spotted you, you might have quickly maneuvered to avoid an encounter. Occasionally, or maybe just once and for all, with your fury mounting, you simply couldn't take it anymore and you might have even struck back.

All these reactions are part of the fight-or-flight response, a catchy duo introduced more than a century ago by the physician and scientist Walter Bradford Cannon. Since then, a third F, "freeze," has been added, and the words have been rearranged according to the typical sequence of our fear responses. When frightened, most of us freeze first, before we decide to flee. When these frightful Fs seem like losing propositions, we can become engulfed by rage and prepare to fight. The fight-or-flight idea caught on, setting the science world on fire, not because of its alliterative catchiness, but because of its profound biological implications. Cannon showed that emotions as different as fear and rage can have the same effects on our bodies, and this equivalence in physiological responses led him to a truly radical hypothesis. No matter how different fear and rage might seem, he postulated, they should share the same anatomical source; no matter how behaviorally different freezing, fighting, and fleeing are, they should be driven by the same internal engine.

Cannon was the chairman of the Department of Physiology at Harvard Medical School from 1906 to 1942. As a medical student, he became interested in the gastrointestinal system and was one of the first to use X-ray technology to generate a moving picture of a bodily function. By stringing together a sequence of rapidly acquired X-rays of the stomach right after the subject finished eating, he generated a movie of peristalsis—that rhythmic churn caused by our stomach muscles that propels food forward. Later, as a newly tenured professor at the turn of the twentieth century, and with the job security that tenure gives, Cannon decided to tackle a topic that might have been considered career suicide for a biologist at that time. He set out to study emotions, those ill-defined mental states that were largely left to students of psychology.

He realized that emotions can influence peristalsis, which seemed to freeze with fright in some of his most fearful subjects. If you have ever felt bloating and appetite loss when stressed, you have experienced the way fear can freeze your gut muscles, bringing your stomach churn to a halt. He also knew that fear can freeze the secretions needed for digestion. Think, for instance, of that dry, chalky mouth you experience right before speaking in public. This fear response is so automatic and uncontrollable that it was used as one of the first known lie detectors. In ancient India, a group of criminal suspects who had been rounded up would each be ordered to chew a spoonful of rice and spit it out onto a leaf; the driest wad fingered the fearful criminal.

Cannon's genius was in engineering how to precisely test the ways different emotional states can influence our bodily responses, by focusing on the body response he knew best: peristalsis. Adrenaline, so called because it was found to be released by the adrenal gland, had recently been discovered. It was the first chemically characterized hormone. (The word "hormone" was coined in 1905—from the Latin *hormē*, meaning, with unintentional relevance to this chapter, "a violent action"—to refer to chemicals, released by the endocrine glands, that arouse or excite.) When Cannon sprinkled adrenaline onto strips of stomach muscles kept alive in a dish, it caused the muscles to freeze, similar to what happens when stress slows peristalsis. These observations not only established that adrenaline is at least one of the hormones by which emotions can regulate peristalsis, but they also led to an experimental paradigm for studying the effects of emotions in a dish. When Cannon sampled blood from the adrenal vein of a fearful cat—one briefly exposed to a dog—and from a relaxed cat, only the fearful blood froze the stomach muscles. But the true shocker came when Cannon replicated the same series of studies with blood he sampled after fear transitioned to rage in a battle-ready cat, which was hissing, baring its teeth, and unsheathing its claws. Rageful blood and fearful blood had the same effect: they both caused stomach muscles to freeze. He went on to show that the contents of the fearful blood and rageful blood coordinate an identical range of other bodily responses—such as increasing blood flow and the energy production of glucose—that seem to make evolutionary sense. Those responses prepare us to confront whatever provoked the fear, in case we have to fight or flee.

Cannon and his followers imagined that the fight-or-flight response should be regulated by the same neuroanatomy. During the first half of the twentieth century, the expedition to locate the source of this response got only as far as the bottom of the brain, or the "brain stem," which contains a structure called the hypothalamus—part of our internal danger-detecting system, the hypothalamic-pituitary-adrenal axis. It was shown, for example, that the hypothalamus regulates the secretion of another hormone, cortisol, which is heavily secreted when we're feeling fear or rage. It turns out that cortisol plays a more central role than adrenaline in directing our bodies' complex responses during fight or flight. So, the hypothalamus regulates the secretion of adrenaline and cortisol, the chemical cocktail in our blood that regulates fight or flight. Might the hypothalamus, then, be the brain's central command of danger management? That seemed improbable. The neurons of the lowly brain stem do not receive the sensory input from the outside world that is needed to detect potentially dangerous signals, nor do they perform computationally complex operations to process those signals in assessing how dangerous any signal really is. It was thus assumed that the expedition to map the brain's source of fear and rage would lead north, to regions higher up in the brain.

Around the same time, researchers found that amygdala lesions, caused by natural accidents or experimental design, eliminated our responses to fear—no freezing, no fighting, no fleeing. Then, in the ensuing decades, nearly all brain regions that engage in danger detection were found to converge onto the amygdala, whose output was found to connect directly with the hypothalamus and other brain stem regions that mediate fear's expression. By the 1970s, it was clear that the amygdala was the brain's central command for danger management. Still, it remained infuriatingly difficult to understand exactly how it functioned in this role. Anatomical studies showed that the amygdala was like an archipelago of separate nuclei. It was assumed that some of these nuclei received incoming information, others discerned which stimuli warranted a response, and still others delivered the commands to the brain stem to initiate freeze, fight, or flight.

Studies that tried to determine the functioning of each nucleus and of the amygdala as a whole were confusing and inconsistent. Imagine spilling out the contents of a box from Ikea, say a dresser. Now imagine trying to put it together without an assembly manual. Given enough time (and patience), you probably could figure out which part does what and how they fit together, but only because you would be able to easily measure the furniture's ultimate functionality—

Is it stable? Do the drawers open?—and only because the furniture is relatively easy to assemble and reassemble as you engage in trial and error.

Scientists began making headway into the amygdala's inner workings in the 1980s, when they had an agreed-upon measure of fear in rodents and when they could carefully and reliably control the fear response. In contrast to fight and flight, so hard to measure in a laboratory setting, investigators realized that it would be more useful to rely on the first F, freeze, as their readout of fear. To have control over the freezing response, they developed an experimental paradigm in which animals form memories between a neutral stimulus and a hurtful one, so that the neutral stimulus is enough to cause the freeze response. "Fear conditioning," as this is called, might have happened to you in school, when a bully's face was paired in time with the bully's hurtful actions, so that your fear responses became associated with his face. After you were conditioned, you were likely to freeze, possum-like, when seeing him. Or maybe, à la Cannon, your stomach knotted up and you lost your appetite.

Once this experimental paradigm had been used across many laboratories, we had a diagram of how the amygdala's individual nuclei connect, showing us which amygdala nucleus does what. Some nuclei are engaged in danger detection and danger analysis, while others trigger our freezing, fighting, or fleeing response. Nearly a century after Cannon proposed his idea that there should be one anatomical source that acts as the brain's central engine of fear and rage, these studies validated his assumption. Moreover, they provided an engineer's blueprint for how the amygdala engine is put together.

While not necessarily its original intent, the research paradigm has taught us a lot about the amygdala and fear memories. For example, we now know that fear memories are formed and stored in the amygdala, and that the more fear memories we acquire, the more active our amygdalae become. The latest study, completed only in the early 2000s, identified the specific amygdala nuclei where fear memories are stored and precisely how this happens. I can now, at last, tell you what happened to your amygdala back in your school days.

Your fearful memory of the bully was formed in your amygdala during a time-locked convergence of inputs from different nuclei. One input was from a nucleus that received information from the visual cortex that coded the bully's face, and the other was from brain areas that coded the pain incurred by his behavior. This synchronized convergence, with hippocampal help, activated the memory toolbox in the nucleus where fear memories are stored, causing the nuclei's dendritic spines to proliferate and stabilize. Now, seeing the bully across the schoolyard was all it took for your amygdala to become hyperactive and for it to drive your brain stem to initiate your fear response—freezing, fleeing, or even shifting to rage that one day when you fought back.

This type of dendritic spine growth helps explain the MRI findings that chimps have larger amygdalae than bonobos. Psychology has shown that fear and rage can transition from one state to the other, and sociology has shown that fear and rage, opposite sides of the same coin, can be transmitted from one person to another. Rage in one person provokes fear in another, and if this fear transitions to rage, it might in turn provoke fear in the first person, perpetuating a truly vicious cycle. This macabre dance and its crushing social consequences are sadly familiar to many of us in the dysfunctional and embattled couples or families we may know.

A scar is formed by the growth of new tissue after an injury. By that metric, chimps' large amygdalae might be considered the result of emotional brain scarring that has occurred while living in a state of perpetual fear and rage. That the amygdala is the region of chimps' brains that differs the most from that of bonobos' brains is itself evidence of how ruthless and unforgiving chimp society must be—and, allowing for some anthropomorphizing, how much pain chimps must suffer, the many humiliations they must endure. One can't help but be bonobo-like and feel sympathy for the chimps' rageful and lonely leaders as much as for those leaders' fearful and cowering minions.

Cannon incorporated evolutionary biology into his original framework to propose that the emotional twins fear and rage are acquired traits not only within an individual but also within a species, via evolutionary adaptation and natural selection. These traits are, as Charles Darwin said, "born of innumerable injuries in the course of evolution." Just as an individual will learn to become more fearful depending on their early life memories, so too will a species evolve to become more fearful in particularly scary and dangerous environments.

Which brings us back to bonobos and their distinguishing prosocial traits: altruism, compassion, empathy, kindness, tolerance, playfulness, loveliness, even sexiness. When such "multi-trait syndromes" occur, Nature often selects for one key trait, and the others come along as a package deal. Evolutionary biologists have convincingly argued that chimps' key adaptive trait is fear, while bonobos' is the opposite, fearlessness. It makes sense that an individual whose social temperament leans fearful and rageful will be less inclined to be altruistic, compassionate, empathetic, kind, tolerant, and playful. This fearful difference in social temperament is the key that drives the rest of the antisocial or prosocial traits, with each personality a better fit for a species' divergent environment. The chimp, for example, competing for limited resources with the bigger and stronger gorilla, lives in a much rougher neighborhood than the bonobo.

Canine behavior exemplifies the theory that fear is the one defining trait that can account for other tag-along social traits. It is generally agreed that dogs diverged from wolves when the first proto-dogs were, presumably as an accident of Nature, endowed with the bravery required to approach human settlements and benefit from the ample food source we call garbage. Out of this key fearlessness trait, all the other friendly traits that we associate with dogs emerged. In a long-term study meant to recapitulate this evolutionary process, investigators selected foxes for reduced fearful aggression and allowed them to breed, repeating the cycle for more than twenty generations. Not only were the ultimate progeny less fearful and less aggressive than their forebears, but they also demonstrated a range of other dog-like pro-social traits. They formed closer and more intimate social bonds, not only with foxes in their own pack but also with other foxes, even with animals of other species; that is, they did not suffer from that dreadful condition "fear of strangers" (xenophobia). Their temperaments were more playful, and they generally seemed to enjoy life more, even wagging their tails. And, to Cannon's posthumous delight, their glandular secretions had ebbed, with the hormonal thunderstorm that orchestrates fear and rage eased to a drizzle.

Much has been clarified about the amygdala in just the past decade. While the degree of our amygdala activity is likely determined in part by our genes, we now know that it strongly depends on the fear memories we acquire. We also know that the amygdala is the anatomical engine that drives both fear and rage, emotions that can influence our social temperaments. We can plausibly hypothesize, therefore, that relaxing our amygdalae would in turn improve our personalities.

One way to forget a fearful memory and to relax our amygdalae is by relying on the mechanisms discussed in chapter 1. Memories are formed and stored in amygdala neurons using the same memory toolbox that induces dendritic spine growth in all neurons. Fearful memories, like all memories, are flexible and can be sculpted using the same forgetting toolbox that causes spine shrinkage. Imagine encountering your now much-picked-on bully, but only after he has received years of psychotherapy or spent months in an ashram seeking and attaining spiritual enlightenment. He has overcome his rage issues and has become, if not exactly a sweetheart, at least pleasantly benign. An electrode placed in one of your amygdalae would show high neuronal activity upon seeing him for the first time after his transformation. But upon seeing him over and over again, this hyperactivity would gradually abate as the spines that stored the original fearful memory shrink and your amygdala neurons start the slow process of fear forgetting.

While patients who successfully complete exposure therapy report an improvement in their social dispositions, this process of fear forgetting is thought to take days or much longer if not experienced continuously. Because of this protracted period, it is hard to verify how powerfully an improved social temperament is linked to fear forgetting per se. In the interim, many other factors might have contributed to the patients' improved behavior.

Thanks to all those tinkering studies that mapped the amygdala's internal circuitry, we now know how to rapidly induce fear forgetting. If the amygdala is considered to be an engine, certain amygdala nuclei have been identified that function like engine pedals. One amygdala nucleus functions like a brake, and drugs have been found that effectively press the brake pedal, inducing a rapid slowdown in amygdala activity. Another amygdala nucleus is more like an accelerator, and drugs can ease up on that pedal to decelerate the amygdala. When such drugs have been given to animals, they have both had a similar effect: rapid fear forgetting that is directly mediated by a reduction in amygdala activity.

It turns out that many drugs that we have been taking for decades—whether by prescription or recreationally—work in part by pressing down or letting up on these amygdala pedals; we just didn't know it. Many of us have unknowingly experienced the social bliss that comes with a reduction in the kind of amygdala activity that causes fear forgetting. If you have ever noticed how that first alcoholic drink (less so the third or fourth) improves your feelings toward others, you have likely experienced the way low doses of alcohol decelerate amygdala activity. The same is true for those who have taken prescription drugs such as benzodiazepines (e.g., Xanax or Ativan) or nonbenzodiazepine derivatives (e.g., Ambien or Lunesta). These drugs are collectively categorized as anxiolytics because they reduce the anxiety and dread caused by our fear memories. Next time you drink alcohol or take one of these medications, pay attention to how you feel. Do you experience at least an inkling of the many prosocial traits that distinguish bonobos from chimps? Of course, alcohol and anxiolytic medications will sooner or later affect other brain regions, and depending on our different sensitivities, increased dosing will cloud these experiences.

If you've tried the recreational drug MDMA (methylenedioxymethamphetamine), you are more likely to have experienced that constellation of traits with cloudless clarity. MDMA is a complicated drug, and it is impossible to intuit its effects on the brain from its chemistry alone. The testimonials from MDMA consumers are, however, uniform, with the descriptions of its effects overlapping, if not nearly synonymous with, bonobos' distinguishing prosocial traits. A recent brain imaging study provided a clue to MDMA's singular neuropharmacology. The drug was found to reduce brain activity, and the greatest reducing effect was localized to the amygdala and the hippocampus to which it connects—a brain pattern of fear forgetting if there ever was one. The drug's effect can be so rapturous and exhilarating—the word "love" is one of the most common descriptors used—that it is commonly called ecstasy.

So here we are, as in the previous chapter, back to the amygdala, and also back to love. It would seem unlikely that the amygdala evolved its braking system simply for our recreational or bacchanalian pleasures. Clues to the true purpose of the amygdala's inhibiting mechanisms come from oxytocin, a chemical naturally produced in our brains. Oxytocin was first characterized at the turn of the twentieth century, when science was hormone obsessed. A spike in oxytocin was found to occur during childbirth, and its physiological effect was to relax uterine muscles during delivery. Oxytocin was also found to be released during breastfeeding, thereby accelerating the production of milk. But motherhood is obviously more than just delivering and feeding babies. Even an unsentimental cynic would have to concede that a mother's love is so patently obvious as to be an objective truth. And here is where oxytocin becomes interesting, when its effects are extended from physiology to psychology—from the maternity ward, where it is sometimes used to facilitate childbirth, to maternal love.

Only in subsequent decades did oxytocin's role in the psychology of motherhood begin to emerge, when the degree of maternal bonding was found to be influenced by manipulating oxytocin levels in the brain with artificial injections.

More oxytocin produced more bonding between mother and child. Many other forms of social bonding also have been found to be oxytocin sensitive. The holy bond of matrimony, or at least its secular manifestation, monogamy, is oxytocin sensitive, as is even more casual social bonding. In fact, oxytocin does not have to be artificially administered; the levels of natural oxytocin produced in our brains spike during activities that hook us up both socially and sexually.

Oxytocin is a deceptively simple chemical produced in brain stem nuclei, where it can be released to other brain regions. The amygdala is among the brain regions most sensitive to oxytocin, which behaves like those drugs that hit the amygdala's brakes, slowing down its overall activity. All mammals depend on maternal social bonding and benefit from forming social bonds throughout their families and communities. Mammalian brains have evolved to produce oxytocin. If fear memories help drive our antisocial fear responses, then it's plausible that Nature engineered a braking system into the amygdala engine to countervail our fear memories with fear forgetting in the service of social bonding. Think back to your first day of kindergarten. You were perhaps excited, but certainly at least a little bit apprehensive. In such a situation, our initial fear response might be to momentarily freeze, or even wish to flee; for some of us this fear might uncontrollably transition to aggression. These are legitimate, not neurotic, responses, as any new environment *is* potentially dangerous. The lurking hazards of interacting with new people are real and, especially to a young psyche, genuinely scary. While these fear responses help minimize risk, increasing our physical and psychological safety by causing us to either withdraw or lash out, they impede all stages of social bonding, from that very first interaction to the forming of meaningful and lasting friendships.

The fear memories that influenced exactly how fearful you were on that first day of school began to form when you were a toddler. Even if you were raised in a coddling environment, the emotional thorns of early life had already left their memory marks on your amygdalae by the time you entered kindergarten. Remembering fears is of paramount importance for our survival, and the amygdala is hardwired to learn and hold on to fearful events. Luckily, it is also bestowed with a braking system that can temper those fears and the memory of them, plausibly allowing us to forget them just enough to be able to form and forge social bonds.

Just as stress hormones can initiate a vicious downward spiral of fear and rage between two individuals, oxytocin can trigger an elevating spiral, an uplifting pas de deux. Mere eye contact can cause an upward spiral of oxytocin release. This eye-to-eye communication occurs even in our social bonds with dogs. A recent study showed that when dogs and humans gaze into each other's eyes, their oxytocin levels increase, and when oxytocin is administered artificially, it increases eye locking. Try it yourself: gaze deeply into your dog's eyes, and you will immediately sense a spreading tenderness that almost certainly has something to do with oxytocin release and amygdala relaxation—a restorative effect that helps explain the popularity of dog therapy programs.

Oxytocin has sometimes been called the love hormone, but this is reductionism gone too far. While it is true that oxytocin's first claim to fame was its role in maternal love, it would be wrong to conclude that all forms of oxytocin-sensitive social bonding represent love. Some oxytocin-dependent emotions might meet this highest degree of social bonding, but—perhaps thankfully, given love's searing intensity—not all do. And for those who consider monogamy the paragon of matrimony, be careful what you wish for. Bonobos, the quintessential prosocial personalities, are polygamous. It does seem very likely, however, that forming the mutual trust that defines a social bond—any kind, but particularly love—requires some degree of fear forgetting in order to open our minds and our hearts, and thereby expose ourselves to life's social dangers.

One explosively traumatic event is all it can take to damage our brains, impairing the normal balance between emotional memory and emotional forgetting, and disordering our personalities. Forgetting some of the pain of a traumatically emotional memory can prevent or help heal certain psychopathologies. Normal life experiences can also affect this balance, though more subtly, tilting us in either a prosocial or an antisocial direction.

There is a tendency to pathologize the rageful temperaments of chimps, mob bosses, ruthless politicians, and schoolyard bullies. Extremes in normal traits can at some point cross over into the abnormal—for example, sadness transitioning to pathological depression. The question is where to draw the line between normal and abnormal. Remembering fears serves an obvious purpose. Clearly, it would be wrong to diagnose chimps with amygdala hyperactivity and treat them for it, as their temperaments are well suited for their rough-and-tumble neighborhoods. One criterion for considering a human condition pathological, or at least justifying treatment, is if it causes suffering in a person's life. Society and the legal system can judge those brutish types whose amygdalae photographically remember every slight and every humiliation, and who live in a state of perpetual fear and rage. But doctors should diagnose and provide care only for those who seek it, and resist requests—as I have with growing frequency in our current political landscape—to "armchair diagnose" the morally reprehensible.

For those of us lucky enough to have blissfully forgotten some of our fears—who have the capacity for compassion and can keep rage in check—the neurobiology of forgetting teaches us why we should sympathize with those suffering souls who can't forget and who live in the viciousness of their fear and trembling. We should be thankful that most of us can, if not silence, at least occasionally muffle the scream of our fear memories. Without the benefits of fear forgetting, we would live dreadfully lonely lives.

FIVE

LIGHTENING MINDS

Inscrutably, his answer was "maybe." I was sitting with Jasper Johns, one of the greatest living American artists, in the dining room of his Connecticut home. My wife and I have a farmhouse across the border in New York State, and we had met Jasper socially. Fascinated by the brain, Jasper has on numerous occasions invited me for lunch: to tour his studio, to stroll the sweeping grounds of his country estate, and to chat about, among other things, brain matters.

It was a rare experience to explain to this particular artist, whose most famous work depicts generic objects—flags, numbers, targets—how the visual system represents objects in the visual processing stream. How, step by step, the visual system reconstructs the object: first colors and contours, then individual components in lower-order cortical regions, and finally the snapping together of the unified whole as information converges on higher-order cortical regions in the central hub. How common objects can then network with other information through the binding of multiple cortical hubs. And how emotions can become entangled in this sensory web by binding with subcortical hubs.

Jasper began painting his first in a series of flags in 1954, a year after he completed his service in the U.S. military, during a nationalistically heightened postwar era. Recounting how my comrades and I in the Israeli military had accelerated emotional forgetting by the ironic use of flags, with all their nationalistic associations, I asked Jasper whether he thought his flags might have similar utility—for either his personal well-being or the country's. Notorious for skirting questions about his creative process, he offered only the noncommittal "maybe."

Jasper, who would never be accused of verbosity, is more communicative about art in general and the art of others. Two shareable conversations we had were about creativity and forgetting, one on the pathological variant of forgetting and the other on normal forgetting.

I was initially surprised when Jasper expressed a strong affinity for the Abstract Expressionist Willem de Kooning and his body of work. From what I knew about art history, Jasper and his contemporary Robert Rauschenberg, considered precursors of Pop Art, were largely responsible for bringing an end to the dominance of Abstract Expressionism. De Kooning completed his first in a sequence of canonical "Woman" paintings just two years before Jasper began to paint his flags. Two years but worlds apart in style, content, and, most interesting to me, the pebble-splash associations these paintings provoke: de Kooning's *Woman I*, so expressive with its swirls of color and not-so-abstract depiction, triggering associations that flicker between emotionally charged concepts of mother or lover; and Johns's *Flag*, two-dimensional and rendered with deceptive simplicity (a close-up reveals the complex mix of oils and wax), a dispassionate depiction of a common object eliciting an associative flicker between the quotidian and ironic social commentary.

De Kooning was one of the longest-lived Abstract Expressionists, dying in 1997 at the age of ninety-two. He began his last series of paintings in the 1980s, at a time when he was already manifesting symptoms of dementia. Jasper wondered about the cause of his dementia, and I agreed to perform some medical detective work, informed by our dialogues and the information that was publicly available. Korsakoff's syndrome, which is caused by a deficiency of thiamine and is often associated with excessive alcohol drinking, was apparently considered, given de Kooning's long history of heavy drinking (up until the 1970s, he was still going on benders). So too was vascular disease, which can cause dementia when a spray of strokes strikes cognitive regions. As I discussed in chapter 1, vitamin deficiencies are typically excluded by blood tests, and vascular disease can be excluded by MRI or CT scans. While these clinical studies were not available to me, his treating neurologist diagnosed "probable" Alzheimer's disease as the cause of de Kooning's dementia. Clinicians, particularly those at that time, use the term "probable" after excluding other potential causes of dementia and when, in light of a telltale cognitive profile, they feel confident that Alzheimer's is the culprit. When they are less sure, either because some of the exclusionary tests are positive—for example, the detection of strokes—or because of an ill-fitting cognitive profile, they downgrade their diagnosis to "possible" Alzheimer's disease. I inferred, therefore, particularly since de Kooning's history of heavy drinking was well documented and must have been assessed, that the appropriate blood tests and MRI scans had been ordered and were negative.

I was able to glean the second requirement justifying the probability of Alzheimer's disease—the cognitive profile—from descriptions about de Kooning's behavior years before he was finally diagnosed. One telling anecdote described how de Kooning could not remember a close friend's recent paintings, even though he had just viewed them, but could faithfully recall the artist's paintings from long ago. That is a malfunctioning hippocampus speaking. Or, more expressively, that is a hippocampus "crying for help," borrowing a term from the eighteenth-century pathologist Giovanni Morgagni, one of the founding fathers of modern medicine, who first fully articulated the need to listen for a disease's anatomical source. Approximately six years after that episode with his friend, with his cognition worsening, de Kooning was formally diagnosed with Alzheimer's. Pathological forgetting of recent events followed by chronic deterioration in other cognitive abilities are signposts of the disease's typical anatomical itinerary, which begins its death march in the hippocampus before migrating upward to higher-order cortical regions, including the cortical hubs described in previous chapters. Alzheimer's disease was probably the right diagnosis for de Kooning.

I reported my conclusions to Jasper, with the caveat that

my clinical sleuthing was far from a formal evaluation. While not chatty, Jasper is an exacting conversationalist. In a gentle manner and with a soothing southern lilt, he gets to the bottom of things with simple and concise questions. "Why was the certainty level only 'probable'?" I explained how definitively diagnosing Alzheimer's disease, particularly at that time, requires a microscopic viewing of the brain after death. I described how in that process a neuropathologist examines thinly cut brain slices under a microscope in search of the disease's defining pathologies: disheveled nests of wiry protein fragments, one kind found inside neurons, called neurofibrillary tangles, and the other kind found in between neurons, called amyloid plaques. From visiting his studio, I understood the importance Jasper places on methods and process—on material engineering in the visual arts—and I sensed that he would appreciate the technical details. I told him that no matter how thin the brain slices, the pathologies remain invisible until the slices are treated with special stains. Only then do they glare up into the microscope, obvious and ugly to the eye.

"Why is it called Alzheimer's disease?" Jasper asked me. I explained that the disease is named after Alois Alzheimer, the German neuropathologist who first reported on these pathologies in 1906, in a patient who died with dementia. Until then, dementia was thought to be a form of madness, which implied a certain degree of willful wickedness or moral depravity rather than a true disease. True neurological disease at the time required visualizing pathologies after death. By visualizing the invisible at the beginning of the twentieth century, Dr. Alzheimer not only established that dementia was a neurological disease but also set the stage for all the medical advances in the disease achieved at the century's tail end.

"Surely these pathologies existed before 1906?" Jasper said. Yes, of course they did, but it was only in the late nineteenth century that neuropathologists began soaking brain slices with chemicals that were originally developed as dyes for the German textile industry. Some of these dyes, it turned out, colored not only cloth but also Alzheimer's disease. Jasper's expressive eyes twinkled when he heard this historical nugget, a node of convergence between the visual science that is pathology and the visual arts.

De Kooning's brain was not evaluated after his death. New tools that today can detect indirect evidence of the pathologies in living patients were not available at the time. These Alzheimer's disease "biomarkers" are measured by performing a spinal tap, looking for snippets of amyloid plaques and neurofibrillary tangles flicked off into the spinal fluid, or by injecting patients with radioactive stains that safely bind these pathologies so that their glow can be visualized with imaging cameras. So, no autopsy and no biomarkers. Nevertheless, from the evidence accumulated, I comfortably concluded that the diagnosis of Alzheimer's disease was almost certainly accurate.

The dialogue about de Kooning's cognitive demise continued, and it became apparent that Jasper's interest was not just academic. In 1995, Jasper and a group of art historians were summoned by the executors of de Kooning's estate on

Long Island to evaluate his last series of paintings, completed in the 1980s. Aware of de Kooning's diagnosis, the experts were tasked with determining the quality of the paintings—whether they represented the late style of an aging creative genius and thus should be considered a last but legitimate chapter of his oeuvre, or they were so compromised by his disease that they should be placed outside the scope of his artistic trajectory and shelved from public viewing so as not to damage his legacy. It was my turn now to boil this down to a simple question: Could Alzheimer's disease impair an artist's ability to be creative?

We neurologists commonly field variants of this question when patients, family members, and sometimes even—most awkwardly—the courts ask how the disease can affect career performance. The answer, never easy to ascertain, depends on the stage of the disease. But it's much easier to come up with an answer for someone in a common profession than for a career artist.

Alzheimer's is a gradually progressive disease. While the disease's ground zero has been pinpointed, its time zero, its precise beginning, has never been clocked. We do know that it takes decades for a patient to develop dementia. By tracking large groups of people over decades, we know that Alzheimer's begins in the "preclinical stage," when neurons in a hippocampal region called the entorhinal cortex begin to malfunction, but only subtly so. The afflicted person might notice occasional memory lapses in newly learned information—trouble recalling the name of someone recently met, for example—but these remain subjective and are not reliably detected by formal memory tests. As the disease progresses over the course of many years, patients enter the "prodromal stage." During this middle stage, the disease is still largely confined to the hippocampus, but it begins its blanket killing of neurons, causing consistent and detectable memory impairment—forgetting a movie seen last night, a dinner party attended last weekend. From that point, it typically takes five to ten years for the disease to enter the "dementia stage," as the disease migrates out of the hippocampus up to higher-order cortical regions—hubs of complex networks that are the seats of other cognitive abilities. Most prominently affected during the early phases of dementia are the cortical regions where information is stored and processed and from which it is retrieved: the cortical hubs of the sensory processing streams described in previous chapters. Now, besides just the hippocampus "speaking," the patient begins experiencing overt pathological forgetting: forgetting events from younger years, forgetting the names of friends, forgetting words, forgetting travel routes, forgetting how to return home.

Unfortunately, the disease does not stop there. Ultimately, after many years, during which time the disease is confined to these cognitive areas, it spreads throughout the cortex, robbing a person's personality and personhood. And then it dives deep below the cortex to the nuclei—button-sized clusters of neurons—in the brain stem that are critical for maintaining consciousness and for basic bodily functions, including sleeping, eating, and breathing. It is this end stage of Alzheimer's disease that strikes terror in patients and family members when the diagnosis is invoked. But for most of the disease's course, it is confined to information-processing areas of the brain.

"Just shoot me if I get it!" is the declaration I often hear when chatting about Alzheimer's disease in social situations. I initially thought this was a reaction to knowing about the end stage of the disease and, without getting into the ethical debate over euthanasia, a tenable response. I came to realize that it was more about fear of the prodromal stage and the early phases of dementia, about losing one's cognition. Now that I have viewed Alzheimer's disease up close and throughout its long course, mainly as a physician but also in my own family, I appreciate how wrong this suicidal reaction is. Without minimizing the anguish caused to patients and their families, I must say that none of my patients during the prodromal stage or even the early phases of dementia wish to die. It turns out that we can lose many of our cognitive faculties and still engage with others and enjoy life. For some this might seem obvious, but even in our current age, which places so much importance on information—its processing, storage, and retrieval—my patients have taught me that we tend to overvalue computational dexterity. We seem unaware that many cognitive abilities are not critical for our being—our core personality traits, our ability to socialize with family and friends, our ability to laugh and love, to be moved by beauty. But clearly, cognition is important for most of our careers, certainly for mine, and its gradual demise exacts a price.

While reviewing with Jasper the stages of the disease, a required preamble to deciding whether an Alzheimer's disease patient can practice their profession, I realized that I had shifted into lecture mode. We settled back into a proper dialogue as we entered the more challenging part of the discussion: "staging" de Kooning's disease while he painted his last series, and determining whether the disease might have affected the paintings' quality.

We discussed whether an artist could still create works reflecting his or her genuine creativity if, like the patient H.M. described in chapter 1, the hippocampi were excised from the artist's brain in later life. The artist's cortical visual processing stream, from beginning to end, would be intact, and so would the associations the visual cortical hub had formed with other sensory modalities and with emotions—as long as this binding process had occurred months before the hippocampectomy. Our conclusion, therefore, was that an artist's creative process could survive when he or she was in the preclinical and prodromal stages of the disease.

The problem was that in de Kooning, the disease had very likely spread into the cortex during the 1980s, while he was creating the paintings in question. He was formally diagnosed with Alzheimer's disease dementia in 1989, and based on our experience as neurologists, patients enter the dementia stage a few years before they are diagnosed with dementia. It is hard to be precise because the anatomical progression of Alzheimer's disease is not like a sequential series of region-by-region surgical excisions. The disease's stages are not discretely demarcated, but bleed into one another. Once the disease gradually migrates into a new region, it festers there for years as it sickens neurons before slowly killing them off. Despite the inherent anatomical fuzziness, Jasper and I were able to conclude that for most of the 1980s, de Kooning's disease had already migrated out of his hippocampi and that his visual processing stream was affected—but only its higher-order cortical regions, at or around the visual central hub. I immediately emphasized what seemed very relevant to our brain-mapping mission: that only during the very end stages does the disease begin to propagate down the sensory processing stream, from the central hubs to the lower-order cortical hubs, where colors and contours are processed. There is little doubt that these lower cortical regions in de Kooning's brain were relatively intact during most of the 1980s.

And so, through our dialogues, Jasper and I finally, as best we could, completed an approximate map of de Kooning's disease. De Kooning painted much of his last series after the disease had already affected his visual cortical processing stream, but only its upper reaches. Even there, however, in the central hub, many neurons were still alive, although they were "sick." The complex perceptual processing that occurs in them, therefore, would have been dimmed but not darkened; the links these neurons form with other information and emotions would have been loosened but not absent. In contrast, the neurons in his lower-order cortical regions would have been robustly healthy throughout the decade. This map of disease might help explain, at least neurologically, why de Kooning's style was so dramatically different at that time than it was before. Gone were the complex and emotionally charged depictions of figures, objects, or landscapes, rendered in lushly dense and variegated brushstrokes. These were replaced instead by sparse ribbons, simple colors and contours.

The question was whether the disease map could inform us of the quality of his later paintings; whether his cognitive function was sufficiently impaired to affect his professional capacity as an artist. Neurologists are asked this all the time, in helping decide whether a patient can no longer practice their profession. Neuropsychological tests provide the objective evidence—equivalent to a mechanic's checklist when examining a car—to document when a run-down hippocampus and prefrontal cortex, and other cortical regions to which they connect, impair the ability to, say, process and remember information, to speak fluently, to calculate and manipulate numbers and other abstract symbols, and to navigate in space and time. This cognitive checklist can be translated practically to help decide when most patients' professions are likely impaired, but not so for an artist.

Outside a neurologist's scope, the determination of when an artist's performance is subpar has to be left to other experts—in this case Jasper and the group of art historians who visited de Kooning's estate in 1995. Collectively, they concluded that except for his last few paintings, most were of sufficiently high artistic quality, were consistent with his artistic trajectory, and belonged as legitimate parts of his oeuvre. This approval formed the basis of, and gave the green light to, an exhibition of these paintings, ultimately opening to nearly universal acclaim at the Museum of Modern Art in 1997.

Illuminated by the de Kooning case study, I articulated a

practicing neurologist's conclusion: that artists can, in principle, practice their profession through the preclinical, prodromal, and early dementia stages of the illness. The most interesting conclusion to me as a cognitive scientist was that the creative process, even its genius, can occur when the upper regions of a sensory processing stream are impaired—its rich weave of associative networks in tatters—and when the stream is dominated by lower cortical regions, where sensory processing is simpler and associative networks are sparser and more rudimentary. Jasper's response to this neurologizing was a cocked head and a sly smile.

Jasper did offer glimpses into his creative process for painting his *Flag*. Not to me, but in a number of interviews published in the 1960s, where he acknowledged how sleeping was part of the process, and how the inspiration for painting an American flag came to him in a dream. Dreaming is known to be a fertile state for creativity not just for Jasper and other artists but for scientists as well. I knew, however, that there was no point in my asking Jasper to expand on his quotes, as his coyness on this topic would undoubtedly prevail. But I sensed that he would be interested in new findings on the biological purpose of sleep, which are beginning to explain how dreaming benefits creativity.

The body's need for sleep has remained one of biology's great mysteries. A few minutes a day are all we really need to eat and drink enough for survival, but we are forced to dedicate many hours to sleep, to disconnect from the world around us, with all its lurking dangers. Those of us lucky enough to get the full eight hours our bodies crave, and even those who claim a couple of hours less, end up spending almost a third of our existence exposed and vulnerable to our surroundings. Despite this exposure, alternating wakefulness with sleep is so essential to life that no being with a complex nervous system has managed to escape this daily cycle of existence. Mammals do it (from humans to rodents), vertebrates do it (from fowl to fish), even the lowly invertebrates do it (from flies to worms). Yet, unlike nourishment and hydration, whose necessity for our bodily functions is easy to explain, the need for sleep has remained unknown.

Many hypotheses have been proposed in an attempt to explain why, despite the fact that conscious awareness of our surroundings increases our chances of survival, we are forced to dedicate hours a day to a slumbering oblivion in order to survive. One hypothesis, whose general contours were proposed a quarter of a century ago, has slowly amassed circumstantial support. Only in the past few years, however, with the development of sophisticated technologies, has it been tested and confirmed.

Francis Crick, the scientific luminary who shared the 1962 Nobel Prize in Physiology or Medicine for describing the double-helical structure of DNA, sparking the molecular revolution described in previous chapters, shifted his focus later in his career. Brazenly, he decided to tackle the most intractable questions in the brain sciences—the nature of consciousness and the mystery of sleep. In 1983, he published a theoretical paper that hypothesized about sleep's biological purpose. He summarized his elaborate idea in one pithy and startling conclusion: "We dream in order to forget."

Recall that the neuronal correlates of memory are those small protrusions from dendrites, the dendritic spines. The billions of neurons in our cortex each have thousands of dendritic spines, so the number of individual spines is truly astronomical. Their sole purpose is to modify their size, and the number of neurotransmitter receptors contained within them, with experience. Each spine is endowed with the molecular machinery to sprout in response to experience, and each experience triggers vast fields of spine growth.

Imagine spending a day in your life wearing eyeglasses with a built-in mini-camera documenting, frame by frame, the thousands of images you experience. Viewing your daily odyssey as a slide show later that evening, you would recognize many, if not most, of the experiences. Each moment of recognition would reflect the growth of millions of spines distributed across your cortex. Even though many experiences shared overlapping information and therefore overlapping spines, each was at least partially distinct. Your recognition of them is psychological evidence that your brain must have grown, even if just microscopically, throughout the day. Now imagine a whirlwind tour of the world, a weeklong adventure spent jetting to, and spending a chock-full day of sightseeing in, very different environments—a city, a jungle, the mountains, ancient ruins, the desert, a bucolic countryside, and, most boring, a resort island. Each day, you are pumping your brain with thousands of distinctly vibrant memories, each fragment of memory a lawn's worth of spine growth. Leaving aside the spatial problem—that your rigid skull prevents your brain from significantly expanding in size—such spine growth gone wild would cause cognitive havoc. Each spine can only grow so much, and sooner or later your cortical spines would fill to capacity. When this happened, like a saturated digital picture with no contrast across its pixels, memory snapshots of previous experiences would be whited out and left indistinguishable. Running out of spines, the cortex would eventually have no room left for new memories to form. With your cortical sensory processing regions densely populated with overgrown spines, even your perceptions of the outside world would be affected. They would become distorted as neurons in these cortical regions became overly excitable to incoming information, and they might even be deranged by the information overload, which would throw the normal sensory flow out of whack.

Crick first proposed in 1983 that sleeping solves this problem by what has come to be called "smart forgetting," an idea that has been modified and refined over the years by his students and other investigators. Based on the principles of neuronal plasticity, sleeping, particularly dreaming, should have a dual and opposing effect on the fields of new spines grown in response to our daily experiences. While we dream, the hippocampus stimulates and replays fragments of our experiences in the cortex, but not the full episode experienced in all of its elaborate intricacies. Dreams are like those "previously seen" recaps in TV series, in which only the most important snippets are reshown, the very few needed to capture and reinforce the gist of the story line. In so doing, the hippocampus persistently stimulates a few privileged cortical spines, stabilizing into a memory those few whose growth reflects the gist of our daily experiences. More sweepingly, however, the vast majority of new spines are left unstimulated while dreaming. Unstable and neglected, vast fields of freshly grown spines should, according to the general hypothesis, wilt back down. After a good night's sleep, we might expect to see some pockets of newly grown spines, now stabilized into a memory. But the net effect, comparing the cortex at the day's end and the morning after, would be spine shrinkage—that is, the net effect of sleeping is forgetting. While it is true that a secondary gain of the large-scale pruning that occurs during sleep would benefit memory—topiary-like, accentuating its details—sleeping's main purpose, according to this hypothesis, is to refresh the cortex. By cleaning and clearing the cortical slate, sleep reopens the cortex to accommodate future memories; by reducing neuronal excitability and effectively deleting extraneous cortical information, it preserves the regulated processing and flow of sensory input.

While this hypothesis makes sense, it is only in the past couple of years that studies have empirically validated its key assumption. In 2017, using powerful new microscopes and other sophisticated techniques, researchers were at last able to investigate spine size across large swaths of cortex. The results were strikingly clear: the net effect of sleep is to cause wholesale spine shrinkage—to cause forgetting. To paraphrase one of Crick's former students, who performed many of the seminal studies documenting sleep-induced forgetting, sleep is "the price we pay" for having a nervous system so eager to learn that it evolved trigger-happy spines, ticklishly responsive to our external worlds. The added elegance of this hypothesis is that it explains why we must disconnect from the external world for so long and on a daily basis. Shrinkage of spines cannot happen immediately. It takes hours for the delicate molecular machinery that governs active forgetting to carefully disassemble a newly grown spine. So, in contrast to hunger, which can be sated by a few ravenous mouthfuls, or thirst, which can be quenched by a couple of long swigs, forgetting can't be rushed. Forgetting takes its slow and deliberate time.

The behavioral consequences reported by people who are forced to go for days without sleep provides more experiential support for the hypothesis. If sleep is important for memory, as earlier hypotheses have contended, then sleeplessness should ultimately cause the kind of memory loss seen throughout the stages of Alzheimer's disease. But this is not the case. What is reported instead are symptoms consistent with neurons that are overly excitable to sensory input, with cortical regions in sensory overload and overflow, all to be expected if sleep's primary purpose is to forget, to shrink spines and erase information. The telltale symptoms of this lack of forgetting, which are devastatingly experienced by virtually anyone forced to go days without sleep, are distorted and deranged perceptions. Sleeplessness affects every part of the visual processing stream—distorting the way we see colors and contours and a precept's component parts, and ultimately deranging the combined whole, even fleetingly causing us to hallucinate.

The benefits of sleep on creative insight also come into

play when looking at the effects of sleep-induced forgetting. Psychologists have pored over the introspections of individuals who are generally agreed to be highly creative—visual artists, poets, novelists, musicians, physicists, mathematicians, and exceptional biologists. A unifying thread among these testimonials has emerged. Colloquially, "to create" implies novelty or innovation; "to be creative" suggests a broader generative capacity. But the recurrent theme that epitomizes the creative process is not generating something brand-new out of the blue. Rather, a creative spark occurs when unexpected associations among existing elements are suddenly forged—a sort of cognitive alchemy. Phrases people use to describe creative insight include how elements in one's mind engage in "combinatory play," how they "collided until pairs interlocked . . . making a stable combination," or how they "drew together beneath the surface through the almost chemical affinities of common elements." My favorite description is from the poet Stephen Spender, who described his creative process as "a dim cloud of an idea which . . . must be condensed into a shower of words."

Psychologists set out to devise a behavioral task that captures this creative crucible. Consider the following three words: "elephant," "lapse," "vivid." Think of a fourth word that relates to all three. The answer is "memory." How about a word that is associated with another trio: "rat," "blue," "cottage"? If you answered "cheese," you are right. But even if you did not, take a moment to reflect on these two answers. (See page 208 for more examples.) Once you put the words together and come up with or are shown the right answer, its accuracy is obvious, and you experience an aha moment.

There is no obvious route a mind must take, no formula for how to cognitively compute the right answer. It just happens. The correct answer is always there, somewhere in your cortex. You know that rats eat cheese; you have eaten, or at least seen, blue cheese or cottage cheese. But if you were asked to free-associate to "rat" alone, "cheese" might not come first to your mind. Unless you are a cheesemonger, "blue" might elicit "sky"; "cottage" might spark "house." Only if you are a pest control expert, a ratcatcher who has experimented with various baits, might the word "cheese" come first to your mind. Similarly, only if you are a memory expert like me might the word "memory" be your response to "elephant," "lapse," and "vivid." On the flip side, the strength of my association with words linked to "memory" can potentially constrain my creativity. I cannot, for example, see one of the sea's most fabulous creatures, the seahorse (*hippocampus* in Latin) without being locked in to an immediate association with "memory."

And this is exactly the point. Creativity requires preexisting associations—requires memory—but they must remain loose and playful. The artists' testimonials teach us that creative abilities are forged by immersion in various elements and the establishment of associations between them, but only when the links are relaxed. All visual artists immerse themselves in visions, poets in words, scientists in facts and ideas. But what sets the great ones apart is that their associations are not set in stone.

Loosening associations, relaxing links, associations that are set in clay not in stone: all are required for creativity, and all sound like forms of forgetting. Is this true? Evidence that forgetting is beneficial for creativity first came from studies in which psychologists used various ways to either strengthen or loosen associations between word pairs, like "blue–sky" or "cottage–house." For example, by repeatedly exposing subjects to word pairs, researchers found that they formed tighter memories between those couplets and predictably initially performed worse on the creativity task. Subjects' performance gradually improved over the next few days, however, an improvement that tracked with forgetting's known timeline.

While those findings are interesting, other evidence that links forgetting to creativity comes from sleep studies. These studies clearly show that our creativity, whether measured by the creativity word task or on other measures, significantly benefits from a good night's sleep and in particular from our dreaming. And when examined, this benefit did *not* occur because sleeping is somehow restful. *Nor* did it occur because dreaming happens to sharpen a few memory snippets of what our minds were exposed to throughout our daily peregrinations. Most of the studies were performed before the definitive evidence validated Crick's prediction that we sleep in order to forget much of our quotidian memories. Nevertheless, with the benefit of scientific hindsight, the inescapable conclusion is that we are most creative when associations of what we do remember are kept loose and playful by sleep-induced forgetting.

Anyone can learn why we need to eat, how food is digested, how nutrients are delivered to cells, and how cells combust them for energy production. But nothing can teach this need more acutely than experiencing a bout of starvation. The powerful yearning, the overwhelming craving, you feel for sleep after a long and eventful day is the best way to fully grasp forgetting's need. The blissfulness of a good night's sleep is to experience your dendritic spines neatly trimmed, your mind lightened and refreshed to record the day ahead. The mental static and derangements you feel after nights of insomnia, in part, are the result of a brain overloaded with unnecessary information.

As Jasper and I were wrapping up our long discussions, we considered whether forgetting evolved specifically for creativity. While it is undoubtedly true that we and other species have benefited from creative insight, it is more likely that creativity piggybacked on the forgetting process, whose primary reason to evolve was for the cognitive and emotional benefits described in previous chapters. But it is still true that by lightening our minds, forgetting unmoors us from memories that weigh our minds down and prevent flights of fancy and creativity.

SIX

HUMBLE MINDS

Decision-making in medicine, as in all aspects of life, is vulnerable to our personal biases. Doctors, for example, are asked to refrain from treating family members or even close friends because, according to the American Medical Association's code of ethics, close relationships can "unduly influence" our "professional medical judgments." We can break this ethical code—for example, during a medical emergency—but when doing so we must be fully aware of our biases in order to reduce medical error.

A "minimizing bias" is a common concern I hear from colleagues when treating loved ones. We sometimes tend to minimize symptoms—fail to listen carefully or take them seriously—when expressed by someone we know well. A doctor friend of mine, for example, dismissed his three-year-old daughter's complaints of always feeling thirsty, only to belatedly find out that she had diabetes. Any interpersonal relationship can bias decision-making, and while there is no ethical code against physicians treating physicians, here too we need to be on the lookout for prejudice. If a minimizing bias causes us to do too little or to act too slowly, the opposite, a "maximizing bias," can sometimes occur when treating another doctor. Perhaps fearing the embarrassment of botching a diagnosis for a colleague, we often have a tendency to do too much, such as ordering an excessive number of medical tests.

Awareness of our cognitive selves—our cognitive aptitudes but also our biases and pitfalls—is called metacognition. I experienced a metacognitive moment when I was looking over the list of patients scheduled to see me one afternoon and recognized the name of Dr. X, a world-renowned infectious disease expert from our medical center. I didn't know him personally, but I knew him by reputation and had once referred a family member to him. He, like I, seemed to be aware of the potential biases when treating a fellow physician, and from the get-go he clearly tried to ensure that our physician-physician relationship would not influence our new patient-physician one. Not only did he seem very deliberate in introducing himself and emphatically calling me Dr. Small, but he was not wearing his white coat even though he had rushed over to my office in between seeing his own patients. I followed along, purposefully refraining from engaging in the small talk common among hospital co-workers and not correcting him by insisting he call me Scott, which I routinely do with nearly everyone who is not my patient.

In his late forties, Dr. X presented with an odd chief complaint. He felt that he was more forgetful than others and always had been, and he was interested in knowing whether this was objectively true. When I asked "Why now?" he replied that he had simply become more introspective as he entered midlife, an answer that turned out to be only half-true. In contrast to Karl, my patient with the "steel-trap" mind, Dr. X reported that he'd felt since elementary school that his memory was poorer than that of his peers. I pointed out that his memory could not have been truly terrible, since his résumé demonstrated academic excellence, first at a competitive college and then in medical school, where success is notoriously dependent on memorization. He explained that he was good at "cramming," that he was able to "hang on to" new information for a few days, but then, "like invisible ink," memories would seem to evaporate from his mind. Whether recalling jokes, names of famous actors, or academic information that he transiently mastered—historical dates in college or cranial nerves during medical school—his memory was patently worse than others'. Important for my assessment, he was very clear that his memory had not worsened over time and that this was simply how his "brain was hardwired."

After performing a neurological exam that showed nothing abnormal, and aware of how a maximizing bias can lead to an overexuberant ordering of medical tests, I resisted the temptation to order an MRI or the standard blood work. But to help determine whether there was any objective evidence that Dr. X's memory was subpar, I asked one of my psychological colleagues whether she would be willing to administer our typical battery of neuropsychological tests in formally evaluating a patient's cognition. She agreed. Two weeks later, after the tests were completed and scored, the three of us reviewed the results in the cognitive testing room down the hall from my clinical office. Perhaps to put him at ease—this was a colleague, after all—the psychologist began by first complimenting Dr. X's high IQ. Either because he intuitively knew this or had been told it before, he glossed over the praise and got down to the medical business at hand, asking about his chief complaint, his memory. I explained, as I do with many of my patients and my students, how neurologists parse the vague concept of "memory" into different types organized by the brain's anatomy: the hippocampus and its role in forming new long-term memories; the posterior cortical areas, in which memories are ultimately stored; and the prefrontal cortex, which aids in our retrieval of memories from the cortical storage sites. I explained how each memory system is evaluated by neuropsychological tests, and taking this as her cue, the psychologist proceeded to recite his scores.

His scores on tests that evaluated his memory storage sites and memory retrieval were normal, but his hippocampal function was lower than normal. I quickly jumped in to qualify that although it was lower than average, it was not abnormal; that is, like variations in height, his memory was simply on the shorter side of normal. Knowing his medicine, he continued with the analogy and politely pointed out that while a person's ultimate height is genetically determined, malnutrition can stunt this natural capacity. Might this also be true for memory? I replied that he was in principle correct. Nutrition has long been assumed to influence cognition. For example, in 2014 we identified one group of nutrients, flavanols, discovered to be specifically linked to hippocampal function. Found in many fruits and in green tea, flavanols are found in the highest concentrations in cocoa beans, which is why they are sometimes called "cocoa flavanols." A quick review of Dr. X's diet revealed that it was excellent, and while blood tests are being developed to measure flavanol consumption, they are not yet available. I felt that Dr. X's subnormal memory was most likely written in his genes. Knowing he would be interested, I promised to send him some recent papers that have identified proteins, and the genes that encode them, that are linked to hippocampal-dependent memory, although I explained that this genetic information has no therapeutic implications. I told him that his hunch was likely right, that he was probably hardwired this way. He acknowledged that his father's memory had been famously bad, and so this made sense. He seemed satisfied with this objective confirmation of his poor memory. Knowing thyself, even if discovering that thyself is subnormal on some traits, can be gratifying.

Dr. X then circled back to his IQ, idly wondering how his could be so high with such a poor memory. The psychologist fielded this question. She explained how IQ is measured and what determines the score. The commonly used test she administered is a composite of different subtests, each one measuring different functions of the brain that collectively contribute to our intelligence. Only one depends heavily on our hippocampal system, measuring the extent of our vocabulary and our knowledge of geographical or historical information. Dr. X's scores on these tests were average at best. His extraordinarily high IQ was driven instead by his outstanding performance on the other, less-hippocampal-sensitive subtests. Psychologists often cluster many of these subtests into one cognitive capability, which is referred to as the important-sounding "executive function."

Important-sounding, perhaps. All-important, yes! But executive function is more obscure and difficult to explain than, say, memory or language. Dr. X's academic mind was piqued, and he clearly wanted to hear more. The psychologist and I obliged. Like the top level of a government or corporation, the brain's executive branch is first tasked with reasoning through a stream of incoming information and solving problems presented by this "intel." If, after internal deliberations, the brain's executive decides that corrective actions are needed, its second function is designing and implementing the most rational plan of action. While neuropsychological exams that evaluate the executive's second function do exist—used, for example, when someone is applying for certain high-performance jobs—the neuropsychological tests we use focus primarily on the executive's first function: reasoning and analytical problem solving. It was this part of the exam that Dr. X aced and that was the main determinant of his high IQ. Dr. X understood this right away and boiled it down by saying that this was basically math, a skill he had always excelled at. "But what is math?" the psychologist and I responded nearly simultaneously and with a touch of exasperation. There is nothing wrong with boiling down—we are all reductionists, after all—but saying that reasoning is math is circular, almost like saying that thinking is to think. Psychologists want to understand the mental operations that allow us to reason, mathematically or otherwise, and neurologists want to know the neuroanatomical map of where these operations are performed.

Psychology first. Psychologists have described another kind of memory operation that is needed when we reason or problem solve—when we do math. This one is called working memory. Not to be confused with hippocampal memory, working memory refers to our ability to briefly remember information long enough to manipulate it. It is our mental scratch pad, where we do back-of-the-envelope calculations, our mental juggler that allows us to keep multiple pieces of information simultaneously in the air. Solving that classic, and to many students annoying, school math problem in which two trains are speeding toward each other at different rates and you are asked to calculate their collision time or place, is an example of using working memory. Rotating complex three-dimensional objects in our minds or finding the one item that doesn't match the others in a sequence of items both benefit from this analytical memory system. A simpler example that illustrates working memory's role in mathematical thinking is an exercise we use called serial sevens. Patients are asked to count down from 100 by 7—i.e., 7 from 100 (93), 7 from 93 (86), and so on—and we measure how far down they can go. By no means confined to mathematical thinking, working memory extends to all kinds of information. Another test we use is asking a person to spell the word "world" backward. This test illustrates that while working memory can in principle function independently of our long-term memories—when tasked with purely analytical problems—it can also benefit from them, since remembering how to spell "world" forward eases the pressure on working memory. In all these tests, subjects need to briefly store new information or access old information (numbers, words, objects, concepts) and keep them in their minds long enough to perform an operation on them (calculation, spelling, rotation, comparison). Unlike with hippocampal-dependent memories, once the operation is completed, we can crumple up the information and throw it away.

Neurologically, executive function is headquartered in the prefrontal cortex, a brain area already discussed in the context of memory retrieval. The prefrontal cortex is vast, one of the largest divisions of our brains, and it is organized as a warren of regions that act collectively. The prefrontal cortex receives all the inputs needed for executive function. It receives real-time factual updates about the outside world from the cortical sensory areas, which I discussed in the context of faces and names, and on-the-fly danger assessments from the amygdalae. It can also download old information on demand. Working memory is performed in the part of the prefrontal cortex that is akin to the White House Situation Room, where new and old information is quickly analyzed in order to make decisions and action plans. The prefrontal cortex can also serve the second part of executive function, designing and green-lighting a plan and then deciding how best to mobilize and implement it, by having direct outputs onto all motor areas of the brain.

As always, the neurology of a complex brain function is easiest to grasp in its absence. Executive function is impaired in patients with lesions to the prefrontal cortex, but since Dr. X, like most people, had never seen such a patient, we turned to a more familiar example: the brains and behavior of children. By the time we emerge out of infancy and into childhood, all of the sensory systems in the back of the brain needed to funnel information into the prefrontal cortex are fully developed. All of the motor areas that the prefrontal cortex needs for designing and implementing a plan of action—notably speech and coordinated control of limb movements—also are armed and ready. Even the hippocampal-dependent memory system is operational (starting at around three years of age, which is why we have no memories of our infant years). All that's lacking is the executive's centralized headquarters in the prefrontal cortex. The elaborate construction of the prefrontal cortex lags way behind in the brain's development process, and it becomes fully functional only in our late teens or early twenties. Children can memorize, perceive, and process sensory information as well as adults, but without a mature prefrontal cortex, they cannot reason or decide as well, and they have difficulty controlling their impulses. Neurobiologists know this, but so do legislators, who don't grant children the right to vote, and car insurance companies, which charge teenagers higher rates. Dr. X found all of this fascinating. He thanked us for confirming what he always suspected about his cognitive self and was genuinely grateful for the neurological lesson.

Dr. X's cognitive profile is useful in highlighting another cognitive trait, one that has direct bearings on my Alzheimer's patients. While Dr. X had poor hippocampal-dependent memory, he had good metamemory—that is, he had good awareness of the quality of his memory. A subset of metacognition, metamemory is defined as how well our subjective sense of our memory ability matches an objective measure of it. As it turns out, most of us across the memory spectrum—whether our memory is outstandingly good or bad, or somewhere in between—have pretty good metamemory. Many of our patients, particularly those in the early stages of Alzheimer's disease, maintain normal metamemory: they are fully aware of and acknowledge their gradual, and sadly inexorable, cognitive slide. Some, for reasons we still don't fully understand, are not so fortunate. Not only do they lose their memory, but on top of that they lose their metamemory of that memory loss. And it is these patients, who do not recognize their worsening cognition, who often suffer the most from the deleterious consequences of their cognitive demise.

One of the worst experiences of my professional career was when I was summoned to court to testify against one of my patients, a sixty-eight-year-old man who had totaled two cars as a result of worsening dementia. Because of his poor metamemory, he denied that his cognition had worsened, and he refused to surrender his keys and stop driving. His family felt that they had no recourse but to take him to court, and the judge ultimately adjudicated against him, revoking his license and allowing his family to take away his car. A judicious decision, but a tragedy all around.

I now knew neurologically what Dr. X had always intuited. Among the main anatomical divisions most relevant to cognition, Dr. X's prefrontal cortex functioned above average and his hippocampus below average. He had good metamemory and metacognition more broadly. No other tests were indicated, and as far as I was concerned, his medical case was closed.

But as we left the psychologist's office, Dr. X pulled me aside and asked for a little more of my time. He had something else he wanted to run by me. Since this was not one of my clinical days and another neurologist occupied my office, we decided to walk to my lab in a separate research building. En route, while waiting for the elevator up to my lab on the eighteenth floor, we returned to our physician-physician relationship, commiserating over the chronically slow elevators and gossiping about a rumored change in hospital leadership. Once in my office, Dr. X finally revealed the main reason for his renewed interest in his poor memory. It had to do with medical decision-making.

As a member of one of our medical school's committees tasked with improving the curriculum, Dr. X always kept abreast of the relevant literature. He had recently read a paper on the importance of "intellectual humility" in medical decision-making—for ultimately arriving at the correct diagnosis and treatment plan—and how it was being taught to medical trainees. An extension of metacognition, intellectual humility allows people to be open to the possibility that their initial judgments might be wrong. They are more inclined to change their minds from a first decision to an alternative and more accurate one. Unlike, say, airline pilots, firefighters, car drivers, or perhaps emergency medicine doctors, who need to make fast decisions with immediate and potentially life-threatening consequences, most physicians typically can decide more slowly. When initially evaluating a patient, most of us arrive at a fast and first decision on the most likely diagnosis, keeping in mind a short list of other possibilities. Fast decision-making is often good enough for common and obvious disorders, but for more complicated ones, we usually arrive at the ultimate diagnosis later, after taking time to mull over the problem and, if necessary, review the literature or confer with colleagues. While the ultimate diagnosis is, of course, informed by subsequent tests, good medicine is practiced by first deciding what diagnoses are most likely, so as not to overburden our patients with the discomfort, hazards, and costs of unnecessary tests. Memory is obviously critical for determining what at first appears to be the most likely etiology, or cause of the problem. But if that first hunch is wrong, intellectual humility is ultimately more important, allowing us to change our minds and increasing the odds of our arriving at the correct medical decision. This process is referred to as "truth-tracking," in contrast to "truth-seeking." Most of us seek the truth, but only those with some degree of intellectual humility cognitively track their sometimes slow slog toward it.

There are ways to teach intellectual humility to medical students—by enhancing their awareness of decision-making biases in medicine (those regarding gender and ethnicity are good examples); by sensitizing them to the virtues of changing one's mind; by inoculating them against pride, prejudice, and arrogance—and these are being incorporated into medical education. But Dr. X wanted to know what might naturally cause a person to tend toward intellectual humility.

Universally acknowledged to be an outstanding diagnostician, he humbly submitted that his outstanding diagnostic skills were related to his own high degree of intellectual humility—which he speculated might be attributed to his poor memory and his awareness of it.

He recounted a typical experience in medical school. Clinical training follows the Socratic method. After a team of trainees are led by an "attending," the chief of a medical service, to the bedsides of newly admitted patients, the team huddles in a teaching room or in a far corner of the ward's hallway to review each case. The attending, in full pedagogic mode, queries the trainees for their first diagnostic opinions. There is usually one standout trainee with a superior memory who regularly dazzles by reciting a differential diagnosis of probable etiologies and the medical justifications for the top consideration. Dr. X described this kind of mind as "a machine that sifts through a stack of index cards, each containing a diagnosis and its details, scanning the content of each, until finally finding a match." With some envy but without malice, Dr. X realized that his mind did not work that way, and he was never that person. He could generate a possible diagnosis, but unless he was faced with an obvious case, he remained less sure of what he considered the most likely diagnosis.

One trainee is assigned to each patient and, under the attending's supervision, orders the appropriate diagnostic tests. A few days later, the team reconvenes to determine the final, correct diagnosis. But first, for maximum educational effect, the attending asks the other trainees, those who were not assigned the patient, for their final diagnoses. Dr. X, who had learned not to trust his fast decision-making, would, in between managing the patients assigned to him, take those few days to slowly think about the other cases. It was during those medical school clinical rounds that Dr. X noted that his batting average at getting the ultimate diagnosis right, particularly in complex and challenging cases, seemed higher than those of the trainees with better memories, who were less inclined to change their minds. He began developing a reputation as an exceptional diagnostician—not the first and fastest to diagnose, but in the end often the most accurate. He brought his slow, methodical decision-making to his residency in internal medicine and his postdoctoral fellowship in infectious diseases, and had applied it throughout his career.

Possessed of a remarkably self-aware and reflective mind, Dr. X was familiar with what has been called the "narrative bias": the impulse to make sense of our lives by creating a story line—retroactively connecting events with a sequence of causes—that is either untrue or oversimplified. He was aware that although he took comfort in his own narrative that his poor memory had caused him to improve his decision-making, that story line could be a classic case of a "narrative fallacy." His question to me was simple: Was there any objective evidence to support his hypothesis that a poor memory, which he now understood meant poor hippocampal function, is somehow linked to better decision-making? While I was familiar with the decision-making literature, a topic of inquiry that has become one of the hottest areas in the brain sciences, I admitted that I did not have an immediate answer for him. I promised I would think about the problem and get back to him.

Dr. Daniel Kahneman is by all accounts considered the father of the decision-making field, which was born, seemingly fully formed and out of the blue, in 1974 when he and his close collaborator, Amos Tversky, published their seismic paper in *Science*, "Judgment Under Uncertainty: Heuristics and Biases." This field has been largely co-opted by economics and has proven extremely useful in economic decision-making, where it is sometimes called neuroeconomics. It has been so useful, in fact, that Kahneman was awarded the 2002 Nobel Prize in Economics for "having integrated insights from psychological research into economic science, especially concerning human judgment and decision-making under uncertainty." With the importance of decision-making extending into many other fields, President Barack Obama awarded Kahneman the Presidential Medal of Freedom in 2013.

Besides its groundbreaking science, the classic 1974 paper is also famous for the clarity of its writing. The authors explained what they meant by the terms "heuristics" and "biases," which they applied to cognition, with an analogy from another brain system—perception. We are all familiar with optical illusions—how, for example, a line bookended by outgoing arrows will falsely appear longer than an arrowless line of the same length. Another example is the Necker cube, in which two squares are drawn on a page, offset and connected by lines, and an illusion pops up in our minds: a three-dimensional cube in a two-dimensional drawing. Tversky and Kahneman's paper begins with another common optical illusion, often exploited by artists, in which the sharpness of an object influences its perceived distance: the blurrier the object, the farther away it seems. The primary visual cortex is tasked with quickly deciding the length, volume, and distance of things seen, and it arrives at these decisions by using rules of thumb and computational shortcuts—which is what heuristics are. The computational processing that occurs in the primary visual cortex has evolved to employ heuristics because it is highly probable that, for example, a blurrier object is indeed farther away. *Probably,* but not always. The price we pay for the added processing speed gained by having heuristics built into the visual cortex is that we can be fooled by them. For instance, driving in foggy conditions is dangerous because the visual heuristics can dupe us into perceiving cars as being farther away than they really are. Like any mental trick, optical illusions can delight and have been famously used with fun effect by the artist M. C. Escher, but they introduce potentially dangerous biases in how we see things.

Just as these visual heuristics speed up what we see, Kahneman and Tversky hypothesized, there ought to be cognitive heuristics, mental shortcuts that should speed up our thinking when we are confronted by a cognitive decision—one that requires our executive function. Examples include jurors deciding whether a defendant is guilty and stockbrokers deciding where to invest—similar in kind to a doctor deciding on a correct medical diagnosis. The authors then examined a handful of these cognitive heuristics observed in our deciding minds and showed that just as visual heuristics can bias us to make visual mistakes and experience optical illusions, cognitive heuristics can cause us to think irrationally, decide incorrectly, and even experience cognitive illusions.

Whereas visual heuristics rely on the computational tricks in our visual cortex, cognitive heuristics rely on our memories. To illustrate: What is the result of 5 × 6? Your mind immediately decides on 30 as the correct answer, but not because the working memory in your prefrontal cortex actually performed the slow and tedious calculations of sequentially adding 5 to 5, 5 to 10, 5 to 15, 5 to 20, and 5 to 25. Rather, thanks to your childhood hippocampi, you memorized the answers during elementary school, and your adult prefrontal cortex now simply needs to pluck the memorized multiplication table from your cortical memory site. The ability to quickly recall and cite a memorized table is an example of a cognitive heuristic, a shortcut speeding up your decision-making. Our minds have evolved to use cognitive heuristics because they help us think and make important decisions more quickly. But one of the most interesting psychological insights from this field of inquiry is that even when fast thinking is not necessary—when we are given as much time as we want and thus don't really need to rely on cognitive heuristics—we still prefer using them. The simple reason for this preference is that we are cognitively lazy. We sulkily use our working memories only when we really have to. What is the result of 15 × 16? I can practically hear your collective sigh of mental annoyance at a question that has no easy shortcuts and will take some real mental work.

Our strong preference for using cognitive heuristics can bias us toward arriving at wrong decisions. Try answering this question, developed and introduced in later papers by Kahneman and his group: "A bat and a ball cost $1.10 in total. The bat costs $1 more than the ball. How much does the ball cost?" If you, like the vast majority of people, responded "10 cents," it is because the answer popped into your mind first, and you felt confident enough to give a reply. If you were to slow down your thinking and exert some working memory effort, you would realize that this answer cannot be right, that it is a cognitive illusion. If 10 cents was right, the bat would then have to cost $1.10, and the total cost would then be $1.20. Your working memory will help you realize, slowly but surely, that "5 cents" is the correct answer. But it's what you did *not* say that is the most interesting part of what happened in your mind during this process. Before you replied, many of you, to varying degrees, sensed that something might be wrong with "10 cents" as the answer. So why then, in contrast to 15 × 16—when you would not even hazard a guess, when you immediately realized you needed your working memory—did you decide to go ahead and answer? Unlike 15 × 16, which for most of us is a mathematical question clean of any memory associations, the "bat and ball" question was deliberately constructed to tap into our memories—using familiar objects, familiar units of money, familiar round numbers, familiar transactions. All of these elements were injected into the question like slick oil to force most of us to slip on a cognitive heuristic and say "10 cents."

For the math-averse, before you start thinking that the biases of cognitive heuristics are confined to mathematical decision-making, try answering the following question: Which state capital has taller buildings, New York or Pennsylvania? Most people might jump the gun and incorrectly respond "New York." Even though their slower minds could work to retrieve the memory that New York's capital is Albany, which has fewer taller buildings than Philadelphia, their prefrontal cortices quickly retrieve the louder and noisier New York City from their memory stores. The decision-making field has generated a whole list of questions like this, collectively called the "Moses illusion," because the following question is the one most commonly used to illustrate the point: "How many animals of each kind did Moses take on the Ark?" Most people will immediately say "two," even though on reflection and with slower thinking, many will change their minds and realize that the correct answer is actually "zero." It was Noah, not Moses, who populated his wooden ark with pairs of animals. A person raised with no knowledge of the Old Testament, and therefore no memories of its stories, would not fall prey to this memory shortcut and would have to rely on slow thinking—the use of an encyclopedia or Google—to arrive at the correct answer. When it comes to decision-making of all kinds, it seems, the tortoise mind typically beats the hurried and overconfident hare brain.

The field of decision-making has refuted the long-held explanation for why our cognitive decisions, big and small, are not always purely rational. Until Kahneman and Tversky's 1974 paper, it was believed that our emotions could bias our cognitive decisions. The assumption was that if we could approach a cognitive problem dispassionately with our purely cognitive minds, rational decision-making would prevail. What was so transformative about the paper was that it showed that heuristics are built into our purely cognitive minds. No one, by analogy, would say that we fall prey to optical illusions because of some undue influence by another part of our brains. The same way optical illusions emerge out of the visual cortex's methods of processing visual information, cognitive illusions are embedded in the memory-dependent processing of our cognitive minds, not in our emotions. We all have exceptionally rational friends who seem to make ridiculously irrational decisions. Whether like mental banana peels when the slips are funny or benignly absurd, or like booby traps when they are not, cognitive pitfalls are easier to observe in others than in ourselves, even though we all have the same ones. These pitfalls are not like fingerprints or retinas, unique to each individual. Instead, cognitive heuristics and the biases they cause are generic to our psychological makeup.

Kahneman, therefore, was the best person to answer Dr. X's question about the extent to which poor hippocampal function can play a role in decision-making. A colleague knew Kahneman personally and agreed to put us in touch. I was delighted when he, in response to my formal overture, invited me to visit him in his Greenwich Village home.

His top-floor apartment looked out over NYU. Peering out his windows, I experienced one of those wonderful retrospective sensations where the present unexpectedly closes a loop with the past. It was as an undergraduate in NYU's psychology department that I published my first paper, on how the emotional state of our minds can bias what we see. Back then, I did, of course, know about Kahneman's pioneering work, and now here I was, decades later, about to discuss biases with the master. Insisting that I call him Danny, he politely invited me into his living room, lined floor to ceiling with books, and had us both sit down: me on one large, Cubist sofa, him on an identical and opposing one. Sensitive to his undoubtedly limited time, or perhaps just a little bit nervous, I launched right in, detailing the reasons for my visit. I finally caught my breath, and—in another metacognitive moment, hearing the echoes of my rushed sentences—worried out loud whether I was talking too much. Calmly and with a warm half smile, Danny said, "Well, you are here to talk, aren't you?" Thus we began our first in a series of dialogues—either in his apartment or at one of his favorite neighborhood restaurants—on decision-making, the influence of memory, and what factors go into the changing of our minds.

I came to realize that his first response in putting me at ease was not simply the politeness of a much-sought-after professor who has learned to calm overeager students. Rather, his whole demeanor, even when we debated points of disagreement, was solicitous and reassuring. It occurred to me that Danny, who was in his eighties, had that priestly quality that comes to a scholar who has lived his life plumbing the pitfalls of our minds.

. . .

Indulge me in a thought experiment that will employ the full neurological cast of characters involved in decision-making. Imagine you are a member of a counterterrorism SWAT team. Late one afternoon, you're mobilized to a local college where armed white supremacists are holding students hostage in the library. En route, in the back of the black SWAT van—speeding, sirens blaring—your amygdalae mobilize your hippocampi into quick action as you and your team memorize all the relevant information: the library's layout, its entrances and exits, the estimated number of hostages. Most important, intel has come in with photographs of two of the three suspected terrorists. Upon arrival, you and your team assemble just outside the library, double-check your gear, and quietly crouch on either side of the front door. With the negotiations stalled and the hostages deemed in imminent danger, your commander gives the signal: your team rams open the door, hurls stun grenades inside, and rushes in. Within seconds, you need to scan the library, identify the suspects, and decide whether, whom, and how to shoot. You must distinguish the white supremacists from the hostages and discern the safest way to disarm the former without harming the latter. Managing your amygdala hyperactivity, your blood now spiked with cortisol and adrenaline, is the easy part; this comes from years of practice and experience. The harder part is the decision-making. The hardest part, of course, would be the remorse you would feel if, making the wrong decision, you killed a hostage by mistake—even if it turned out that you did everything right and a formal investigation exonerated you.

One terrorist looks just like his picture. When he refuses

to put down his rifle aimed at your team, you decide to shoot. Near an inner door stands a man you think is the second terrorist, but unlike in his photograph, he now has a full beard. Less certain, you hesitate, but your mind is flexible enough to recognize him from his mugshot and identify the door as one of the library's exits, at which he must be standing guard. Your mind quickly computes both facts, and you feel certain that you have identified the second terrorist. When he refuses to heed your demands to drop his weapon, you shoot. You know there is a third terrorist, but which one is he or she? Now with very little certainty, your mind makes calculated guesses: likely a man, not a woman; likely white, not a person of color. You also know that white supremacists often shave their heads and tend to dress in black leather. Your mind quickly assigns probability as you scan each person, until you isolate someone in the back of the library who you judge must be the third terrorist, just as he is reaching into his leather jacket for what you assume is a pistol. You shoot. It turns out you are wrong. You've mistakenly killed an innocent college student.

Each decision to shoot was processed by the executive function in your prefrontal cortex, where, when a threshold of confidence was surpassed, your decision was green-lighted. The first decision required very little prefrontal cortex effort, since your memory of the terrorist's picture was crisply obvious. The second decision required more deliberation—more prefrontal cortex effort—to weigh two less-certain observations. You thought you recognized the man from his photograph, but you could not be certain because of the suspect's beard. Able to mentally rotate the library's map, your prefrontal cortex reasoned that this suspect was possibly standing guard. On its own, each observation might not have met the confidence threshold needed to pull the trigger, but together they did.

So what led your prefrontal cortex to misjudge and tragically misidentify the third terrorist? The dominant factor was a version of a cognitive heuristic called "preference by association." Without having seen a picture of the third terrorist, you had no actual memory of him or her. Nevertheless, over the years you've formed a network of memory associations of what a white supremacist looks like. When forced to decide, and without certain knowledge, your prefrontal cortex was biased by riding the coattails of this associative network. This particular cognitive heuristic is often exploited by advertisers. When deciding what product to buy, we are subconsciously influenced by the company the product keeps. If you are Photoshop savvy, download two photos of the same car, then drop into one a picture of a Labrador retriever. Flash both photos to your friends and ask them to quickly decide which car they prefer. Biased by the memory network they have previously formed of the adorable, lovable Lab, almost all dog-loving people will choose the car with the dog.

You might recall that when psychologists were first describing the kind of conscious memories that are formed with hippocampal help, they called them "explicit memories." They did so because we are commonly consciously aware of the associative network of explicit memories. For example, if I ask you to remember a childhood friend, you can explicitly and consciously recall their name, their personality traits, and when and where you first met. Preference by association is an example of an implicit memory, because you subconsciously form memory associations between, say, the car and the dog. It is the subliminal quality of implicit memories that is exploited by advertisers, because consumers, unaware of the associations, are less likely to override the biased preference.

There are other ways in which implicit memories can force your decisions. Here's a famous example of how this works. Imagine that I said to you, "Say the first word that comes into your mind when you see 'SO_P.'" If I had surreptitiously made any prior reference to food, that previous memory—implicitly planted and thus outside your awareness—would increase the likelihood of you saying "SOUP." If, instead, I had planted references to cleanliness in your memory, your prefrontal cortex would more likely come up with "SOAP." This "memory priming" might seem obvious, but recall the "bat and ball" question, which was deliberately designed to force you to answer incorrectly because it used familiar objects to tap into your implicit memories.

This psychological dichotomy between explicit and implicit memories also agreed with the conventional neurology of memory. Explicit memories became neurologically defined as memories that depend on the hippocampus for their formation, as we have seen in previous chapters. In contrast, the neurology of implicit memories was defined as involving those that occur completely independent of the hippocampus and its function. According to this view, as a SWAT team member, your first two correct decisions were based on hippocampal-dependent explicit memories and thus relatively easy for your prefrontal cortex to judge. As indicated by the title of Danny's seminal 1974 paper, it is "judgment under uncertainty," however, that is more interesting to researchers in the decision-making field, because it is under uncertainty that our heuristics are more likely to cause us to misjudge and "slip," primarily on hippocampal-independent implicit memories. At least this was the general assumption until it was upended in 2012, when a study that has rapidly entered the pantheon of decision-making research showed otherwise.

Published in *Science,* the same prestigious journal in which Danny first introduced the psychology of cognitive heuristics, the study used functional MRI to map the anatomical biology of the preference by association heuristic. Not only did the study show hippocampal involvement; it also found that the hippocampus actually *drives* the formation of implicit memory associations. Across the spectrum, the higher a subject's hippocampal activity, the more likely they were to be biased by the preference by association heuristic; the lower a subject's hippocampal activity, the less likely they were to jump the decision gun. Corroborated by other subsequent studies, this finding has contributed to a major shift in our understanding. Clearly, both explicit and implicit memories can be hippocampal dependent. We now know that hippocampal function drives many, if not most, of the cognitive heuristics that help our prefrontal cortex judge. Often, this is advantageous: when buying a car, say, it's helpful to explicitly remember the terrain or weather conditions you typically drive in. But the hippocampus can also generate subconscious cognitive bait that biases us to make the wrong decision, like buying a car because you implicitly associate it with the cute dog you saw in an advertisement.

Danny agreed that the findings of the 2012 study imply that someone with poorly functioning hippocampi—Dr. X, for example—might be less likely to take the bait, an interpretation that the study's authors generally agreed with as well. Given that, I asked whether he thought someone with Dr. X's hippocampal profile, someone less confident of their memory, might also generally be more inclined to change their mind from the incorrect to the correct decision. Remember that in the SWAT team example, you benefited from your hippocampal function in judging correctly two-thirds of the time, but you suffered as a result of your hippocampi the third time, when you misjudged with tragic consequences.

Danny agreed that confidence is at the heart of decision-making. The term used in the decision-making field is "cognitive reflection," the process by which we mull over our decisions and determine how confident we are before we act. Danny explained that seeking to understand what goes into this type of metacognition is an outstanding quest and an active area of research in his field, but that having more time is not necessarily the critical issue. Even the fast-thinking mind hesitates, self-reflects, and changes a decision on the fly. Cognitive reflection has been used to explain why a person adept at analytical thinking and mathematics is more likely to answer the "bat and ball" question correctly. The interpretation has always been that self-reflection is based on a highly functioning prefrontal cortex, completely independent of any helpful or biased influences from the hippocampus. Exceptional mathematical minds function like a pocket calculator. For them, deciding that "15 × 16 = 245" is wrong is nearly as easy as deciding that "5 × 6 = 40" is wrong, and they don't have to rely on memory-based cognitive heuristics. For them, no matter how deviously clever the "bat and ball" question is in luring us to slip on our implicit memories, their analytical minds have internal indicators that are more likely to alert them to the possibility that "10 cents" is the wrong answer. Cognitive heuristics might fool us into thinking that our initial decision is right, but a superior prefrontal cortex will be more likely to sense that something is off and override that first impulse. In reviewing Dr. X's cognitive profile, it seemed plausible to Danny that Dr. X's flexible mind, whether fast or slow, was linked more to his superiorly functioning prefrontal cortex, and less to his inferiorly functioning hippocampi.

But in Dr. X's absence, I somehow felt the need to advocate for his hypothesis that it was his subnormal hippocampal function that improved his decision-making. I wondered whether "bat and ball"–type questions, which are so dependent on mathematical skills, were unfairly biased. Among the tools that decision-making scientists use, variants of the Moses illusion seem closer to Dr. X's medical decision-making, since by necessity they rely more on our memories. I noted that I found many papers in the Moses illusion literature that excluded subjects with dyslexia. I surmised that this might be because dyslexic brains, which tend to distrust their own reading skills, might be less likely to immediately trust what they first read. In other words, armed with greater "reading humility," they might be more likely to reread the sentence. In that case, even if they at first slipped on a cognitive heuristic, they were more likely to change their minds. To press my point, I returned to Danny's original 1974 paper, where he begins with optical illusions. If I were to look at an optical illusion without my glasses, so that now everything looks blurry, I might be less likely to rely on and fall prey to my visual cortex heuristics. I would have greater "visual humility" and would be less likely to conclude that a blurry object was farther away. It seemed reasonable to me, therefore, that compared with someone who is confident in their superior hippocampal function, someone doubtful like Dr. X would be more inclined to reflect on their first decisions, to have greater "intellectual humility" and to truth-track more carefully.

Although Danny found this interesting, he pointed out that most studies on decision-making have shown that better functioning of a person's analytical prefrontal cortex correlates with better decisions. The common interpretation has been that it is the strength of our brain's analytical executive branch that contributes most to how well we reflect on our decision-making. But, I pointed out, most studies published before the landmark 2012 *Science* paper completely neglected the hippocampus and whether the strength or weakness of our memories played a role in this reflective process. Since it is implicit memories that are important for decision-making under uncertainty, and since by definition we are unaware of our implicit memories, they were considered irrelevant to our self-reflections. Now that we know that stronger hippocampal function—whether the memories it helps form are explicit or implicit—makes us more likely to trip on cognitive heuristics, it has become important to include measures of how well the hippocampus is functioning in next-generation studies. Only future studies that measure both prefrontal cortex and hippocampal function will be able to answer Dr. X's motivating question: whether people with poorer hippocampal function are better at decision-making. Danny reiterated that the investigation of how we self-reflect on our decision-making is precisely where the field is moving.

Perhaps Dr. X's specific cognitive profile, with an outstandingly good prefrontal cortex and outstandingly bad hippocampi, was perfectly calibrated for medical decision-making. Informed by the many studies suggesting that measures of the prefrontal cortex have been linked to better decision-making, Danny thought that it was Dr. X's superior prefrontal cortex that dominated. But just because there is a dearth of studies assessing the hippocampus in decision-making does not mean it plays no role.

As Danny and I were wrapping up, we more generally discussed the idea of intellectual humility. Danny agreed that our cognitive profiles, particularly our relative deficiencies, can temper intellectual overconfidence or arrogance, personality traits that are certainly detrimental to truth-seeking and truth-tracking. He did, however, quibble with the use of the term "humility" in describing someone who is more inclined to change their mind. Nothing gives him more pleasure, he claimed, than changing his mind when warranted, and I conceded that I felt the same way. We agreed that since changing our minds, when justifiable, gives us pleasure, labeling it with virtuous-sounding words like "humility" runs the risk of paying a false compliment. We thought that the word "doubt" would be more suitable than the morally charged "humility." Whether the precise and split-second clockwork of decision-making benefits from a doubtful mind or not, a *doubting* mind is certainly helpful for attaining the ultimate truth.

With some slyness, since at this point I felt more comfortable, I pointed out that we both might be biased by our areas of expertise. Danny's background, and the background of many of his closest early colleagues, is in mathematics, and thus I wondered if he was more inclined to assign a greater importance to the prefrontal cortex and its analytical capabilities in decision-making. I, by contrast, given my "hippocampal centricity," was more inclined to see the hippocampus as the dominant player. With characteristic aplomb, Danny didn't take the bait.

As promised, at the end of my truth-seeking adventure I reached out to Dr. X, and over coffee I summarized what I had learned, including that Danny felt it was Dr. X's highly functioning prefrontal cortex that dominated in enhancing his decision-making abilities. I qualified that conclusion by adding that absent the appropriately designed study, it remains empirically unknown whether Dr. X's hypothesis—that poor hippocampal function is linked to better decision-making—is wrong, and that he should take solace in the fact that his question—simple and straightforward as it might have seemed to him—is currently at the field's cutting edge.

When I mentioned that both Danny and I felt that the

term "intellectual humility" might be misplaced and that "intellectual doubt" might be more appropriate, Dr. X disagreed. He raised the analogous debate over altruism—how some have argued that because people might derive pleasure from philanthropic behavior, it is not necessarily virtuous. Dr. X sided with the view that whatever the secondary gains of selfless behavior, it is the actions in and of themselves that matter, that render them virtuous, and he felt the same about being intellectually humble. I could not disagree.

SEVEN

COMMUNAL MINDS

Walking out to our clinic's waiting room one afternoon, I introduced myself to Joan, an eighty-four-year-old retired schoolteacher who lives in Ohio and whose daughter had scheduled a dementia evaluation for her. Joan was calmly sitting alone next to an empty chair stacked with manila folders and introduced herself in a soft, steady voice.

I initially worried that perhaps Joan had come to the appointment alone. As experts in Alzheimer's disease and related disorders, we typically see patients to provide a second or even a third opinion, so our clinical staff generally operates under the assumption that the patient might have dementia and asks that a family member or close friend accompany them to their clinical visits. If the disease has already spread out of the hippocampus and into the cortical memory stores, we often rely on companions to fill in historical gaps about the patient's premorbid cognition, early cognitive symptoms, and ability to live independently. Assistance might also be needed for patients to find their way to our medical center—in the upper and outer reaches of Manhattan—or navigate the complex maze of buildings to locate the Neurological Institute of New York, where our clinical offices are housed. The day before an appointment, a staff member places a previsit call to the patient and their companion to review how to find us and remind them to bring all relevant medical records.

I politely inquired whether Joan was alone, and she replied that her daughter, Barbara, was with her. She pointed toward the reception desk while saying this, indicating that Barbara was the person having a lively discussion with my staff—apparently on how to improve our hospital's parking. My staff later told me that Barbara had beat them to the punch and called two days before the appointment, seemingly to remind *us* about it, and needed no navigational advice. She had already printed out the Memory Disorders Center's map of buildings and the institute's floor-by-floor building diagram.

It became evident that Barbara, an analyst in a Manhattan financial firm, possessed outstanding organizational skills, which made her not only the perfect patient historian but also a dream daughter. She grew up in Dayton, Ohio, where her mother, Joan, worked as a beloved elementary school teacher until retiring in her early sixties, shortly before her husband's death. When Barbara and her younger brother moved out, Joan insisted on staying in her large family home, which she maintained in immaculate condition with very little help. When they were home for Thanksgiving the year Joan was seventy-eight, Barbara and her brother noticed subtle changes that, in retrospect, Barbara considered the first inklings of her mother's subsequent cognitive slide. She had forgotten to purchase chestnuts, a key ingredient in her family-famous turkey stuffing, and she'd also left a pile of unpaid utility bills on the kitchen table, both the scatter and the outstanding bills atypical for fastidious Joan. Her children dismissed the lapses, attributing them to her focus on holiday preparations and excitement over gathering the entire family, now an extended brood that included children-in-law and grandchildren.

But Joan's cognitive symptoms gradually started mounting. Barbara, who spoke with her mother twice a week, began noticing alarming memory irregularities. Joan would forget that she'd attended a weekly "girlfriends" luncheon the day before and that one of her granddaughters was about to graduate from high school. One Sunday, a parish friend called Barbara to report that Joan had gotten lost on the drive to the church she'd been attending for years.

At this point, Barbara flew home and arranged for Joan to be evaluated by a primary care physician, who ordered a few blood tests, and then by a local neurologist, who ordered an MRI. Joan was diagnosed with "mild cognitive impairment" and started taking Alzheimer's medications. When these had no discernible effects and Joan's cognition continued to worsen, Barbara arranged to fly her mother to New York for an evaluation in our center. The stack of folders I first saw on the chair next to Joan contained copies of not only these most recent tests but also tests and clinical evaluations from decades past, from Joan's primary care physician, her gynecologist, and an orthopedic surgeon who'd treated her for a sprained ankle. The more, the better from our perspective, and they were easy for Barbara to bring, since she kept copies of all of Joan's healthcare evaluations in her Manhattan home, neatly organized by year.

The previous doctors had ordered all the correct blood tests, which were negative, and the MRI was of very high quality. In a zoomed-out mode, no strokes, bleeds, tumors, or other structural lesions were visible. A close-up of both hippocampi revealed that they were smaller than I would have expected, a suggestive but not diagnostic sign of Alzheimer's disease, and an in-office cognitive evaluation implicated the hippocampi as the primary anatomical source. The only missing component of a complete dementia evaluation was neuropsychological testing, which we managed to schedule during Joan's New York visit.

The neuropsychological tests confirmed that Joan's main deficits were localized to the hippocampi. Moreover, unlike Karl from chapter 1, her hippocampus function was profoundly impaired. The neuropsychological tests also revealed subtler defects suggesting that her cortical central hubs, where older memories are stored, had begun to malfunction. Joan's disease had begun its death march out of the hippocampi.

The clinical diagnosis of Alzheimer's disease in Joan's straightforward case, with its telling history and unambiguous tests, could have been arrived at by one of the neurology residents or medical students I teach—and perhaps, after reading this book, by you. The expertise at our medical center is put to greater use when evaluating more complex cases or rarer causes of cognitive decline, and when—in the near future—a new generation of Alzheimer's drugs will require a more nuanced understanding of which drug is best suited to which patient.

I said as much to Barbara and Joan, suggested some minor changes in Joan's medical regimen, and told them that unless there was a sudden change, there was no need for Joan to make the trip back to New York to see me. I would be happy to work with her local neurologist in following her care from afar. I spent most of the hour as a compassionate educator rather than a physician tasked with diagnosing and medicating. I explained our level of diagnostic uncertainty but added that because all the pieces fit so well together, I would not for now recommend any additional invasive tests. I explained that Joan's memory lapses were not her fault, which Joan, who had blamed herself, was relieved to hear. I was candid about the current crop of drugs, which should be considered only first-generation attempts. The benefits are modest at best, but we nevertheless consider trying them in patients because they are generally safe, and because some patients respond better than others. I would consider discontinuing them if Joan experienced any side effects, even subtle ones like vivid and frightening dreams. Without provoking false optimism, and recognizing that Barbara in particular was interested in medical research, I explained how we are finally understanding the disease's root cause, and how, working together with the pharmaceutical industry, we have legitimate optimism for the next generation of truly meaningful drugs.

Joan, ever the caring mother, asked what having the disease might mean for her children and grandchildren. Alzheimer's disease typically strikes in later life, and because most of our patients are parents and grandparents, this is one of the most common questions asked. It is, however, the one that sometimes takes the longest to answer. "Deterministic" genes that *cause* a disease differ from "probabilistic" ones that *influence the risk* of getting a disease. A balance scale, I have found, can illustrate this difference. Deterministic genes contain a mutation heavy enough to tilt your carefully calibrated health scale pathologically off-balance. Risk genes are those whose glitches may be feathery light. By themselves, they do not necessarily mean you will get the disease, but if they are weighed down with other risk genes and risk factors—for example, conditions like obesity, heart disease, or diabetes—they can, as a group, tilt the balance. I explained how these genetic distinctions split Alzheimer's disease into two variants. The first is caused by a single deterministic genetic mutation and is exceedingly rare, occurring in approximately 1 percent of all cases. Because the disease in this case typically manifests in people in their thirties, forties, and fifties, it is often called early-onset Alzheimer's disease. The second variant, late-onset Alzheimer's disease, is the much more common variant, typically beginning in a person's sixties or later. It is sometimes still anachronistically referred to as "sporadic" because its etiology is complex. Family lineage can matter in late-onset Alzheimer's, but only in the sense that our inherited genes influence our risk, or odds, of being afflicted. Because of Joan's age, and because there was no clear family pattern to suggest that she had inherited a deterministic gene from one of her parents, I told Barbara and Joan that she almost certainly had sporadic late-onset Alzheimer's disease. Even if her children inherited one of the probabilistic genes, I explained, they were only risk genes, thereby unburdening Joan from the concern that she might somehow transmit Alzheimer's disease like an infection to her progeny.

We ended up talking a lot about what, for now, is the most meaningful intervention: not pharmacological treatments, but psychosocial modifications. The pathological forgetting that occurs in the early stages of Alzheimer's is not by itself harmful, but locking oneself out on a cold winter's night, forgetting to take one's medications, or mismanaging one's finances might be. By "psychosocial modifications" we mean implementing changes in a patient's life that will safeguard them against the potential harmful consequences of pathological forgetting. In the earliest stages of the disease, these can be as subtle as memory aids, like pillboxes in which medications are allocated according to the day of the week, or asking caring family members to help with bills or finances. Home healthcare assistants might be hired as the disease progresses. Ultimately, a decision will need to be made—ideally involving all family members, as this is often an excruciating process—on whether the patient needs to be moved out of their home and into an assisted-living community.

While Joan never needed to see me again, Barbara would call me every six months, nearly to the day, with follow-ups and updates, and she would send me an annual Christmas card with a merry photo of Joan surrounded by her extended family. As expected, Joan continued to decline according to the typical slow metronome that ticks along for someone with Alzheimer's disease. In our phone calls, Barbara, while always polite, would maintain her cool, almost businesslike demeanor in describing Joan's progression—from her agreeing to daily visits by healthcare attendants as she began forgetting to take her daily medications to very reluctantly giving up driving as she increasingly got lost. Finally, a few years after my evaluation, Joan acquiesced to what she said she would never do. On Easter that year, surrounded by her family, Joan realized that it was time to sell her home—her family's home, and all of their collective memories contained within it—and move into a nearby assisted-living condominium.

Seven years later, Barbara scheduled an in-person appointment with me. I feared the worst, especially when I found Barbara pensively sitting by herself in my waiting room. I was relieved to hear that Joan was doing well in community living, attending many activities and, while slower than before, remaining upbeat. Nevertheless, Barbara was rattled. During their last few visits, Joan had begun having difficulty remembering Barbara's name. It became apparent that Barbara's motive to see me was not as her mother's exemplary caretaker, but rather for personal consolation. Since the suffering caused by Alzheimer's disease is sometimes worse for family members than for the patient, consoling is central to what an Alzheimer's doctor does. Barbara's voice, for the first time in any of our years-long communications, quavered. She wanted to understand—actually demanded an explanation for—how it was possible for a mother to forget the name of her firstborn, a daughter whose relationship with her mother was more friendship than filial; who, when the time came, switched roles and became her mother's mother. Up to that point, Barbara's managerial mind had allowed her to focus exclusively on Joan and approach her disease with corporate efficiency: how best to diagnose it, how best to manage it, how best to care for Joan in this latest and perhaps last chapter of her life. Up to that point, Joan's increased dependence on Barbara had only strengthened the mother-daughter bond. Now, for the first time, Barbara began considering her own needs. She expressed anguish at the prospect of losing a mutually caring friendship with her mother, as Joan's forgetting Barbara's name seemed to imply.

I asked Barbara a number of strategic questions whose answers I felt sure I already knew. Had Joan forgotten other details about Barbara? Yes; for example, Joan would occasionally need reminding that Barbara no longer lived in Dayton, and had difficulty remembering details of Barbara's New York life. Did Joan seem to recognize Barbara at all? Yes; when Barbara walked into the room, Joan immediately looked up to greet her with a sparkle in her eyes and a warm, broad smile. Intuiting where I was going with this, Barbara insisted that there was something nevertheless special—something particularly telling and hurtful—about Joan's forgetting her daughter's name.

In broad strokes, I explained how the hippocampus helps us weave memories of people we know. How elements of those memories are knitted together across a brain network, creating a memory tapestry that is particularly rich with factual and emotional details of those we most care about, and that can, as a whole, be reactivated even if individual nodes in the network malfunction. From what Barbara had told me, I could plausibly infer that while Joan's memory of her daughter had patches worn away by the disease, the network was still intact, and Joan still recognized and certainly still cared for her daughter.

This explanation offered Barbara a bit of solace as it allayed some of her worst fears, at least temporarily. But the subtext was that at some point, as more and more of a patient's memory network drops out, forgetting a name will indicate global network malfunction, that there will be a point where a mother will no longer recognize or care about her daughter. I agreed with Barbara that one of the cruelest consequences of the disease is that as family members need to care more and more for the patient, the patient can sometimes gradually stop caring about them. Many diseases are terrible, but this harsh subversion of the normal caring dynamic distinguishes dementing illnesses like Alzheimer's disease.

From a strictly computational perspective, many elements of a memory network of people we know are equally important, but for most of us and those we care about, Barbara was right: there appears to be something uniquely upsetting about forgetting a name. Even if Karl, my patient from chapter 1, could recognize his new client's face, remember where they'd met, or list facts about her profession or family, the embarrassment of forgetting her name wouldn't have been mitigated. Somehow forgetting a name, more than other personal details, implies that you simply do not care. Dale Carnegie incorporated this psychological truth in *How to Win Friends and Influence People,* one of the bestselling self-help books of all time. Remembering names was one of Carnegie's cardinal rules for succeeding in the business game of life. By forcing ourselves to remember a new person's name, we are conveying—sometimes falsely advertising—that we care enough to remember.

Religions that have no clear concept of an afterlife assign a particular importance to keeping the memory of one's name alive through glorious deeds. For ancient Greeks this form of communal memory was called *kleos*. While I am no longer a practitioner of the religious faith I was born into, I know that remembering names also looms large in Judaism. Most people are familiar with Yad Vashem, the World Holocaust Remembrance Center in Jerusalem. *Yad vashem* is Hebrew for "a monument and a name," a term that comes from the Bible. God commanded that a *yad vashem*—essentially a monument of names—be constructed within the temple walls to commemorate his childless followers so that despite not siring any offspring, they might live on and be remembered for eternity. The cultural importance of remembering names runs so deep in Judaism that it has seeped into its religious vernacular. As righteous men, rabbis are prohibited from using profanities. But they do have one—and as I recall from my yeshiva days, not infrequently used—curse word: *yemach shmo.* Although it is emitted as one long guttural sound, it actually comes from a series of words, again from the Bible. The most damning curse God can cast on an enemy is *ye ma-chek she-mo,* which can be translated as "to blot out his name." Having one's name forgotten, having it expunged from communal memory, is apparently the bleakest fate.

For most people and in many cultures, remembering a name is the ultimate act of respect, while forgetting it is a form of emotional neglect. Caring so little as to forget a name is what, even if only subconsciously, so devastated Barbara when her mother forgot hers, and for those of us with healthy brains, the degree to which we care about another person *does* influence how well we remember their name.

Caring about other people is at the core of ethical behavior. Moral philosophers sometimes contrast ethics and morality based on the intimacy of our caring. In this view, morality is a fixed and universal code of proper conduct toward everyone in the world, including the nameless and faceless unknown. Ethics is more about how we feel and behave toward those we personally know—our family, friends, and community. Morality depends less on memory, in the sense that we would like to believe that proper universal conduct is somehow inborn. Ethics, by contrast, depends heavily on the synthetic hippocampal-dependent memory system. We need our hippocampi to join our amygdalae and cortex in establishing a network of intimate associations: a face associated with a name and other personal details that are then emotionally infused.

Think back to patient H.M., our go-to patient for all things hippocampal. With both hippocampi removed, H.M. still acted morally, and even if he had not, his surgical excisions could not have been invoked as a defense. But he could have been accused of acting unethically. Always courteous, he nevertheless seemed to care not one iota about his doctor, who cared for him for decades. He apparently cared so little that he never bothered to remember her name. Not once did he ask about her personal or professional well-being. More than excising his ability to form new conscious memories, removing his hippocampi stole H.M.'s ability to establish new ethical relationships.

Caring depends on memory, and caring is at the heart of ethics: two simple truths that have justified a philosophical deliberation on the ethics of memory. Is there any ethical benefit to forgetting? Most philosophers have focused on one: the benefit of forgiving. Whether it is forgiving a family member, a friend, or even a larger community, most psychologists and sociologists agree that forgiving requires a certain amount of letting go of seething resentments, of humiliations, of pain. "Letting go" is just another of the many everyday terms that can be neurologically translated into forgetting—in this case, the brain's emotional forgetting that dulls the memory shards of an aggrieved pain.

Dulling shards is not just a "forgetting for forgiving" metaphor for Eric Kandel. Born in 1929 to an assimilated Jewish Austrian family, Eric was raised in a small Vienna apartment located above his father's toy store. After Germany's annexation of Austria in March 1938 and Kristallnacht—named for the shards of glass resulting from the Nazis' paramilitary rioting against Jewish-owned businesses—eight months later, he and his family immigrated to Brooklyn, New York. Eric went on to become a psychiatrist and a leading memory researcher, for which he was awarded the Nobel Prize in Physiology or Medicine in 2000. On the heels of his Nobel fame, he initially rejected overtures from the mayor of Vienna, who wanted his city to somehow vicariously bask in the glory of its native son. Although Eric was originally in no mood for this degree of forgiveness, the mayor persevered and arranged a series of formal dialogues with him. Upon Eric's request, the mayor agreed to implement a number of plans to aid the healing process, placing Eric on a pathway toward forgiveness. While he never forgot the atrocities committed, he ultimately forgave the offending nation—just enough so that he could accept honorary citizenship in 2008. The ethics of this kind of communal forgiveness of an offending nation can be philosophically debated, but in my view it is justified.

At ninety-one, Eric still runs one of the largest and most productive labs at Columbia University, a living testament that memory need not dramatically decline in everyone, and he's been a treasured, longtime academic mentor to me. He agreed to chat with me about the evolution of his relationship with Vienna, and discussing "forgetting for forgiving" with the world's leading memory expert was fascinating. Today, Eric is not just an honorary citizen of Vienna—a city that brutally uprooted his childhood and stripped his family of their dignity and livelihood—but also an active participant in the city's academic affairs and cultural events.

For this degree of forgiveness, multiple action plans were needed, organized around two forgetting mechanisms. The first was the establishment of an annual symposium in Vienna on Austria's response to Nazism. The goal of this symposium was to simultaneously memorialize factual truths and foster reconciliation, both of which require some emotional letting go. In a communal dialogue, the victims of Nazism learn about the perpetrators' warped logic and distorted motivations, and, more important, the offenders learn the extent of the victims' suffering. One goal of the symposium is for the criminal historical episode to be remembered and never forgotten. But another goal of publicly airing the harsh truth is reconciliation—collectively reshaping the memory into the nation's communal consciousness and offering some social amnesty. This recuperative process requires some emotional forgetting. A communal memory, one that emerges from a collection of personal memories, needs to be flexible. If personal memory flexibility requires, as we have seen, active forgetting, then so does communal memory. Amnesty, from the Greek *amnestia,* is by definition a form of forgetfulness.

The second Viennese action plan utilized a more straightforward forgetting mechanism. As our brains intrinsically know, not everything we store is worth remembering, and there is a real advantage—for the sake of our own sanity, or in this case the sanity of a nation—to forgetting details of the world we temporarily encode. Eric knew of a street in Vienna that had been named in honor of a previous mayor, a man who was such a virulent anti-Semite that Hitler cited him in *Mein Kampf*. Upon Eric's request, the street was renamed in 2012. *Yemach shmo!* The blotting out of a street name, it should be noted, is not blotting out from history books the name of a man who should live on in infamy, a man who was not just a casual bigot, but whose vile racism contributed to one of the greatest moral crimes in history.

There is another, less obvious ethical benefit of striking the right balance between memory and forgetting. If memory is required for caring about others, we should consider the ethical consequences of too much memory, which might cause us to care too much.

Honor thy father and mother. Love thy neighbor—or, more accurately from the original text, love thy *friend*—as thyself. Pledge allegiance to the flag. These decrees represent the expanding concentric circles of our ethical relationships, varying in their degree of intimacy from our families to our friends and neighbors to our nation. Among these examples, it is our caring for our nation, our patriotism, that relies most on the hippocampal-dependent memory system in binding facts with emotions. Compared with caring for family or friends, caring for a nation is less innate, more abstract, and thus depends more on learning and memory. It might seem instinctual to say "I would die to save my child" or "I would take a bullet for a friend." But for a country? For this, or even less-extreme forms of patriotism, we need to consciously memorize not only our shared geography and history but also our nation's past glories and agonies.

Let's consider fourteen-year-old Clara, who repeatedly cried out "I want to go home!" as her parents huddled with a doctor around her sickbed. They were vacationing in a seaside resort in northern Spain, away from their home in Interlaken, the bucolic village in the Swiss Alps. The day before, Clara had hit her head while on a sailing trip arranged by the resort's beach staff. Back onshore, with Clara complaining of a headache and some nausea, the resort doctor was called in. He suspected a mild concussion but nothing worse, and recommended bed rest and hydration.

The following morning, besides a minor lingering headache, Clara's nausea had subsided and her neurological exam was generally normal. But she now seemed obsessed with memories of home. This was more than just the normal pining for home we sometimes experience when we fall sick on the road or when we've been away too long. She even stopped eating and drinking because she missed Swiss food so much that she expressed disgust over the "foreign food," as well as scorn for the country's "foreign manners" and the "strange conversation" of the restaurant staff. Just the smell of sea air and the lapping of the waves, so pleasant the day before, now repulsed Clara, who found them intolerably different from the verdant Swiss mountains, alive with the sound of cowbells. With Clara beset by her memories of the "sweet fatherland" and its beloved "national habits," the doctor, consulted again, described this "melancholic delirium" as an extreme bout of homesickness.

"Clara" is actually a composite of a series of young Swiss patients described by Dr. Johannes Hofer in 1688, in his University of Basel medical dissertation. (While many modernizing embellishments have been added to Clara's fictionalized story, Clara's symptoms typify Hofer's cases, and the quotations come directly from his dissertation.) In diagnosing this new disorder, Hofer toyed with a number of new medical terms to describe what he had observed in many of his patients—all Swiss youths, all suffering something akin to extreme homesickness, but medically a lot worse. Two of the medical terms he considered—“nosomania” and “nostalgia”—come from the ancient Greek word *nostos,* which, as used by Homer, conveys a blissful sense of returning home. In the first term, “mania” is from the Greek word meaning “to be mad”; in the second, “algia” comes from *algos,* meaning “pain.” A third option was the less mellifluous “philopatridomania,” essentially meaning “manic love for the fatherland.” Hofer provided no compelling reason for why he settled on “nostalgia,” and in fact, having read his dissertation through modern, medically trained eyes, I think “nosomania” seems more appropriate.

In neurology, we sometimes distinguish between diseases that cause a “loss of function” and those that cause a “gain of function.” Alzheimer’s disease is an example of a loss-of-function disease because when it sickens our hippocampal neurons, it dampens their normal synaptic activity. This dampening of the neurons’ normal firing rate causes us to lose our normal memory function. Gain-of-function diseases result in the opposite. By overstimulating neurons, they prompt the neurons’ synapses to fire too rapidly and the regions of the brain affected to abnormally function too much. Seizure disorders are the most obvious of these “brain on fire” diseases because they occur so suddenly. If the seizure focus occurs in sensory cortical regions, the sufferer can experience a false smell, sight, or sound—all gain-of-function symptoms. If the seizure focus occurs at the cortical central hubs, where memories are stored, the abnormal gain of function stimulates a false memory, causing déjà vu. We now know that hallucinations, delusions, and even obsessions are gain-of-function symptoms caused by a hyperactive brain, although in these cases they are more the result of a slow burn than of a conflagration. The mental derangement that occurs after many sleepless days and the accumulation of too many unneeded memories, as previously described, is another example of a toxic gain of brain function.

Hofer clearly construed nostalgia as a gain-of-function neurological illness—a brain on fire with too many memories—and he located the source of the blaze in a part of the brain where the cortex stores the memory of "home sweet home." Not knowing much about the brain's functional organization (in fact, little was known at the time), he threw an anatomical dart that landed somewhere in the "middle brain." Centuries before anything was known about neurons, synapses, or dendritic spines, he poetically postulated that nostalgia, a disorder of too much memory, was caused by the "continuous vibration of animal spirits through those fibers in the middle brain, where traces of the ideas of the Fatherland still cling." He was physiologically astute in suggesting that this memory fire can kindle and spread throughout the brain. Whereas we might describe an epileptic seizure as beginning in one locus and then spreading throughout the brain to cause a grand mal seizure, he described a nostalgic fire beginning in the "Fatherland" region and spreading via "paths" comprised of "pores and tubes" to cause an all-encompassing obsession, an "afflicted imagination" of the "Fatherland."

As we have seen, PTSD is a toxic gain of emotional memory, whose flashback symptoms can be considered a state of emotional hypermnesia (too much memory). According to Hofer's formulation, nostalgia works similarly. As Hofer explained, because of their own form of hypermnesia, nostalgic patients are no longer able to "forget their mother's milk," and any sight or sound that is vaguely reminiscent of home's sweetness wistfully recalls "the charm of the Fatherland." He concluded by saying that a "meditation only of the Fatherland," by which he meant an obsession, can lead to "stupidity of the mind—attending to hardly other [than] an idea of the Fatherland." And before criticizing Hofer for his insensitivity in his use of "stupidity," we must realize that many such words considered derogatory today were used as formal neurological diagnoses until the late nineteenth century. Adults who behaved like children, for example, were diagnosed as "idiots," and adults who behaved like teenagers were diagnosed as "morons." Hofer was perceptive when implying that an obsession can represent a toxic gain of memory. We now know that cortical areas involved in memory retrieval are indeed hyperactive and hyperconnected in patients with obsessive-compulsive disorder—those who, like Clara, have abnormal and recurring thoughts that detrimentally affect their behavior. For those patients, exposure therapy, which exploits our forgetting mechanisms, is still one of the most effective interventions.

Based on the demographics of his nostalgic patients, Hofer entertained three possible etiologies, or predisposing factors. The first was age. He implied that the impressionability and sentimentality of youth somehow predisposed one to nostalgia. Second, he postulated that some kind of previous childhood "injury" might increase disease risk. Such an injury might, with some modern spin, somehow slow the normal developmental process, preventing our kitschy childish tastes and preferences to fully mature. Finally, since all of his patients were Swiss, he invoked nationalism, and wondered whether there was something special about the beloved "Helvetian Nation" that might explain why, compared with other "tribes of Europe," the Swiss were predisposed to nostalgia.

All disorders begin and are fleshed out from an insightful clinician's hunch. Not every clinical hunch, however, turns out to be a disorder. Unlike Leo Kanner's clinical intuition with autism, Hofer's turned out to be wrong. Nostalgia is not a disorder, a brain on fire with too many memories of one's homeland. But it is nevertheless helpful to use this imaginary illness as a metaphor in our efforts to appreciate the benefits of forgetting for our ethical behavior.

While Hofer's "nostalgia" is not a disease, the term has survived, if not in our medical textbooks, then in our cultural lexicon—largely because the concept of nostalgia was quickly co-opted by the Romantic poets, philosophers, and political scientists, who not long after Hofer published his dissertation began codifying the modern concept of nationalism. By romanticizing homesickness—our caring for our nations, for our motherlands and fatherlands—nationalism was placed on an equal ethical footing with caring for our mothers and fathers.

According to *Merriam-Webster's Collegiate Dictionary*, nostalgia is "a wistful or excessively sentimental yearning for return to or of some past period or irrecoverable condition." There is nothing necessarily wrong with this kind of yearning, and a longing for a paradise lost appears to be part of the melancholic human condition, one that is as old as Adam and Eve. Every nation has its own form of nostalgia, and the only interesting observation is that each believes that its own version is somehow special, despite this generic tug at all people's national heartstrings. It was not ethically wrong for Clara to lovingly remember and pine for her homeland, but once her memory became all-consuming and spread like a moronic inferno, it caused her to rapidly lose her ethical IQ. This perversion, from an ethical love of those we know to an immoral collective hatred of those we don't, is the potential danger of remembering too much, and it applies to all spheres of our ethical lives. Balancing memory with forgetting can help prevent our minds—or, as Hofer put it, our "imaginations"—to be thus afflicted.

All of us, to different degrees, have experienced patriotism, and most of us are justified in caring about our countries, if ethically done. I observed the full range of American patriotism on 9/11, when I witnessed the World Trade Center collapse into a rubble of steel and bones. I was already hard at work that morning, reviewing my lab's budget with our administrators in their offices, which happen to be on the top floor of our medical center's tallest building. The center is situated not only on the northern tip of Manhattan but also on the island's highest point, Washington Heights, so named because it was here that George Washington made a stand against the British at the Battle of Fort Washington. The penthouse office has south-facing windows overlooking the glorious whole of Manhattan, and on that pristine morning we looked up in horror at the smoke billowing out of the North Tower at the island's southern tip. When it became apparent that this was the result of a foreign attack, it was not lost on some of us that it was the first one on the island in more than two hundred years, when our nation was established around the principles of liberal nationalism.

While the twin towers were still burning, we were instructed not to race downtown, which as healthcare providers was our natural inclination, but to stay put, holding down the medical fortress in Washington Heights and awaiting the anticipated stream of ambulances that would be carrying wounded victims to us. With very few survivors of the attack, none came. Idly waiting, eyes glued to the TV as that damned day unfolded, I witnessed a nationalistic discussion evolve among my shocked colleagues. I considered this a healthy remembrance, since I often felt that, compared with my Israeli friends, many of my American ones seemed less emotionally engaged with their homeland. Not that I was critical of that; in fact, I was often jealous that nearly all of my American friends did not have to serve in the military and that they lived in a country whose existence they took for granted. But under these circumstances, with the American homeland under attack, it seemed appropriate to hear them experience a patriotic spike.

Once the towers fell, it seemed as if the activity in Hofer's imaginary brain region started to heat up for most of those huddled in our waiting room. There was a collective sense of wanting to avenge our nation. One potential benefit of growing up in the war-ridden Middle East is that it can sensitize you to the pitfalls of nationalism. Most of the people in that room were unfamiliar with the carnage caused by foreign terrorists, in this case in both their homeland and their hometown, and so this reaction was understandable. What was worrying, however, was when some people's brains seemed to enter an abnormal state of hyperactivity. That toxic gain of function expressed itself, even in some of my most liberal colleagues, in the rageful xenophobia that crept into their discourse, as they seemed to seethe with hatred against all "Arabs," against a whole people. It occurred to me that Hofer was right: this kind of brain on fire, this toxic gain of an all-consuming homeland hypermnesia, caused in some of my very smart friends a temporary moral stupidity.

A few days later, cooler minds prevailed. Undoubtedly, this cooling entailed a very complex process, but in retrospect I am sure that some of the forgetting mechanisms described in previous chapters were at play. We have seen how the forgetting chisel can sculpt away at our memories. We have seen how the therapeutic benefits of emotional forgetting can begin soon after a traumatic event, and that engaging in communal activities can accelerate the process, thus preventing an emotional memory from burning too hot and causing psychopathologies. The same is true for communal memories and social pathologies. For my colleagues and me, this occurred when we joined thousands of American patriots of all ethnicities spontaneously gathering on Manhattan street corners in crowded candlelit vigils; and when we visited the makeshift downtown galleries of the dead and missing, viewing the hundreds of multicultural faces and silently mouthing their names.

Hofer was wrong about nostalgia, but it is not inconceivable that just like there is a collection of cortical hubs that represent a person, there is a collection of cortical hubs that represent our "homeland." Had I been able to record the activity in those homeland cortical hubs over the course of our shifting nationalistic responses and print the record on a long strip of paper—the kind used by cardiologists to record electrical activity from the heart or neurologists from the brain—I would have had a commemorative record of much of what is discussed in this book: a record of how there needs to be a balance between memory and forgetting to maintain a healthy mind, in this case exemplified by our communal ethics.

The electrical record would begin with subnormal spike activity, maybe even a flat line reflective of an ethically dubious state of national forgetting, of caring too little. It would then be followed by a healthy flutter of spike activity, stimulated by a patriotic reminder reactivating our homeland memories back to normal, to where we should remember and care about our country's safety and well-being. But then the homeland memory sparks a wildfire of activity that spreads throughout our brains, a stupefying seizure whose focal source is the homeland cortical hub. Finally, this brain on fire is cooled as the homeland activity dampens—a cooling that undoubtedly has many factors, but at least in part involves normal forgetting, which in this case helps restore most of us back to ethical health.

EPILOGUE

PATHOLOGICAL FORGETTING

"I congratulate you, Dr. Small, on your anatomical skills, but what's the cause?" If this irony-infused statement does not sound vaguely familiar, you have demonstrated that the hippocampus is, often thankfully, not a steel trap and that memories are not necessarily lifelong cortical scars. If it does ring a bell, you might remember that Karl, my patient from chapter 1, asked me this question after I had localized his age-related memory decline to the hippocampus. A subtle—or, with fetchingly quarrelsome Karl, maybe not so subtle—"So what?" was implied by his backhanded compliment. What he wanted to know was not where but why.

You might also remember that I am saddened by not being able to share with Karl the new science of forgetting, most of which has emerged in the ten years since his death. Karl personified the pervasive worry over the normal forgetting we live with most of our lives. Allaying this misconstrued anxiety motivated me to write this book. I set out to share emerging ideas from within the sometimes cloistered halls of academia about the benefits of normal forgetting versus its pathological form, the type that accelerates our forgetting from its baseline state and about which fears are entirely justified.

Karl grasped the premise of anatomical biology for diagnostic purposes—the idea that different disorders target different regions of the brain, and that by localizing which brain regions are affected physicians can diagnose more accurately. I'll conclude this book-long response to the misguided worries over normal forgetting embodied by Karl, by sharing new research on the very legitimate concerns over forgetting's pathological form.

What patients really want to know, what we all want to know, is not just why pathological forgetting occurs, but also how to fix what's wrong, how to cure it. Defective proteins are the primary drivers of disease, and many effective therapies end up trying to correct protein defects in one way or another. The brain is comprised of hundreds of different brain regions, each with its own "neuronal population," each population containing subtly different proteins. The promise of anatomical biology is that if we can pinpoint a neuronal population that acts as the anatomical source of a brain disorder, we can find which proteins within it are defective. "Listening to the cry" for the anatomical source of a disease is a poetic turn of phrase that was used in the late eighteenth century at the dawn of modern medicine, articulating the biomedical search-and-rescue logic of anatomical biology: zeroing in on the anatomical source of a disease promises to reveal its root cause and ultimately to isolate therapeutic clues to its potential cure.

The search for the causes and cures of late-life pathological forgetting had a slow start, lagging behind transformative advances in other medical disciplines. A dominant reason for this slowness—one I explain to my patients and their families to excuse our ignorance—was our confusion over classification. Despite Dr. Alois Alzheimer's describing the disorder in 1906, Alzheimer's disease was shockingly neglected for most of the twentieth century. Dr. Alzheimer's seminal observation was the presence of amyloid plaques and neurofibrillary tangles in the brains of patients who developed dementia in their presenile years, "senility" being a medical term meaning "later life," somewhat arbitrarily defined as beginning in our mid-sixties. "Presenile dementia" is exceedingly rare. So rare, in fact, that while Dr. Alzheimer's findings were considered breaking news because they showed that dementia was a biological disorder and not a patient's willful fault, "Alzheimer's disease" was rarely if ever mentioned in medical textbooks until the late 1970s. "Senile dementia," a progressive cognitive demise that commonly occurs in later life, was long known and exponentially growing in prevalence as medical advances allowed more people to live longer. Nevertheless, it was thought to represent the tail end of the normal aging process, not a disease. Neuronal loss in the brain's memory areas was thought to be part of the normal wear and tear of aging, just like the shriveling of our skin or the whitening of our hair. But as life expectancy went up and more old brains began coming to autopsy, it suddenly dawned on researchers in the 1970s that the very same plaques and tangles that Dr. Alzheimer had observed in presenility were present in senility. The inescapable conclusion was that the two disorders are one and the same. This was a watershed event in the history of medicine. Alzheimer's disease, a diagnosis that now incorporates both presenile and senile dementia, was no longer considered a rare disorder, but one of the most common and dreaded diseases of our era.

The pendulum overswung, however, and before long anyone experiencing mild worsening in hippocampal-dependent memory as they aged—which is to say, sooner or later all of us—was thought to have the earliest stages of Alzheimer's disease. For a minority of neurologists, this made no sense. I, for example, with my background in working with animal models, knew that all mammalian species develop hippocampal-dependent memory decline as part of the normal aging process, and so I found it implausible that humans would somehow be the only mammals spared this normal effect of aging. Those of us in the minority argued that age-related hippocampal dysfunction occurs via two processes—one reflecting aging, the other disease—and noted that unlike, say, presbyopia (normal age-related loss of sight), many people live into their eighties and nineties without developing Alzheimer's disease. The majority camp countered by pointing out that the prevalence of the disease increases with advancing age and if everyone was able to live long enough, everyone would eventually develop Alzheimer's disease.

When I started my own lab in 1998, my colleagues and I wondered if we might resolve this seemingly irreconcilable debate through the logic of anatomical biology. At the time, it was well known that the hippocampus is comprised of a handful of different populations of neurons that are anatomically clustered in different regions. In 2001, we published our hypothesis postulating that while Alzheimer's disease is clearly one cause of age-related hippocampal dysfunction, normal aging must exist as a second cause; and assuming that the two pathologies have different causes, each should target different neuronal populations of the hippocampus. The hypothesis was simple, but testing it was not, since Alzheimer's disease begins by causing neurons to sicken many years before they die off, and the same is true for normal aging. To pinpoint ground zero of Alzheimer's disease versus aging, we needed a camera that could generate a "neuronal sickness" map of the hippocampus in patients who are in the earliest, preclinical stages of the disease.

Functional MRI (fMRI) cameras could, in principle, detect neuronal sickness by mapping how much energy a brain region consumes. Functional MRI generates a sort of heat map of energy consumption, and sick neurons can be either hotter than normal—as we have seen in epilepsy, PTSD, and the imaginary illness nostalgia—or cold with dysfunction, which is what happens in Alzheimer's disease and aging. But the fMRI cameras that existed at the time had a spatial resolution problem. Like an imperfect satellite that can capture a whole archipelago but not distinguish its individual islands, the cameras could visualize only the hippocampus, not its individual regions. Our lab was, therefore, forced to dedicate its first five years to technical research and development. By fits and starts, we set out to improve fMRI so that neuronal sickness could be detected in each region of the hippocampus. With no certainty of success, this was a bit nerve-racking—I was in an early and fragile stage of my career, after all—but the long days and often sleepless nights proved well worth the effort. Our technical innovation succeeded, and once our new and improved fMRI camera was optimized, our hypothesis was rapidly validated.

The advantage of resolving a biomedical debate via anatomical biology is that a picture can speak a thousand words. Once the neuronal sickness fMRI maps were generated in the correct groups of patients, the confirmation of the hypothesis was literally easy to see.

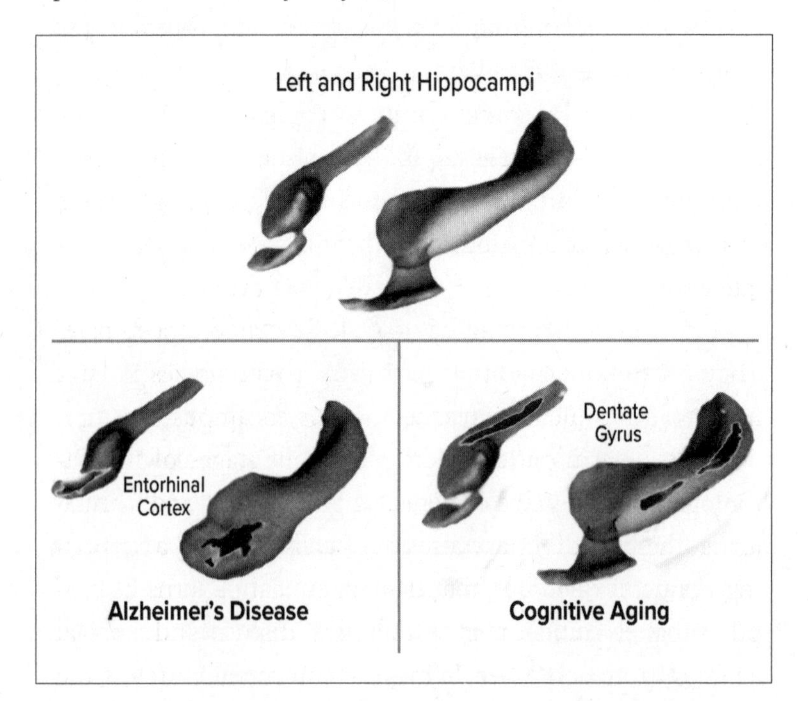

The upper panel in this figure shows the left and right hippocampi. While the panel may seem sculpturally beautiful (it does to me!), it is not an artist's rendering. This is the real thing, extracted from fMRI scans generated from our research participants. The precise anatomy of the hippocampus, with all its curves and curlicues, can be seen in exquisite detail because the images were generated with high spatial resolution. Remember, however, that these images are "functional," not "structural," scans: they contain within them information on which parts of the hippocampi have abnormal energy consumption—that is, which neurons are sick.

In the lower left panel of the figure, which comes from one of our published studies investigating Alzheimer's disease, the neuronal sickness is stained with color. The sickness is clustered in a single neuron population housed in a hippocampal region called the entorhinal cortex.

The lower right panel shows which part of the hippocampi gradually sickens as we age normally. As in Alzheimer's disease, neuronal sickness due to normal aging clusters in a single neuronal island, but a different one called the dentate gyrus.

These and other imaging studies have ended the debate. They have established that two distinct pathologies in later life impair our memory teacher, our hippocampus.

Performed in patients across different stages of the pathologies and in animal models of Alzheimer's and normal aging, these studies have also collectively revealed a surprising bonus: it turns out that the anatomical patterns of hippocampal dysfunction in Alzheimer's disease and normal aging mirror each other. While the entorhinal cortex is the hippocampal region most sensitive to Alzheimer's disease, the dentate gyrus is the region most resistant to it, even as the disease spreads. In mirroring contrast, while the dentate gyrus is the region that is most sensitive to normal aging, the entorhinal cortex is the region most resistant to its wear and tear, even in people in their eighties or older. The mirror imaging of Alzheimer's disease versus normal aging—or a "double dissociation," as this sort of rare anatomical dissociation is called—was not required to confirm our hypothesis, but it strengthens the hypothesis's veracity.

Today, when a neurologist sees a patient like Karl, who presented to me with age-related memory decline, they must consider, and try to sort through, symptoms and test results to evaluate its two possible causes. The resolution of the etiological ambiguity can also help the medical field find the cause of normal age-related memory decline on the one hand, and Alzheimer's disease on the other.

Proteins within cells are the molecular causes of disease. So then: What proteins malfunction in Alzheimer's patients' entorhinal cortex but not their dentate gyrus? What proteins malfunction in the dentate gyrus as a result of normal aging but not in the entorhinal cortex? If our lab's first phase was dedicated to using innovative fMRI tools to establish this anatomical double dissociation, its second phase focused on exploiting this dissociation to identify these rogue proteins. And here too technical innovation was needed, in this case molecular tools that could simultaneously assess the thousands of different proteins and protein precursors that each neuronal population contains. We set out on a molecular mining expedition, after carefully microdissecting the entorhinal cortices and dentate gyri from the brains of older people who died with or without Alzheimer's disease, as well as from the healthy brains of people who died at different ages across the life span.

As we hypothesized, the two populations were found to contain different protein abnormalities that best explained why different hippocampal regions were affected in normal aging and Alzheimer's disease. The isolated defective proteins found in both normal aging and Alzheimer's are tools in the memory's molecular toolbox, which makes biological sense (although confirming common sense in biology is always a bit surprising and tremendously gratifying). In normal aging, the isolated proteins are components of the tools that Eric Kandel and his colleagues first described as the memory toolbox's "on" switch, when information is deemed worth remembering. In Alzheimer's disease, the defective proteins belong to a different set of tools, which work to stabilize new and fragile memories by girding newly grown dendritic spines with their receptors.

I can almost hear Karl: "I congratulate you, Dr. Small, on your molecular skills, but what's the cure?" My answer to him would be a version of the answer I give in both my professional and my public lectures, and that is "Hold on a minute." Finding defective proteins from a neuronal population that is emitting the cry of a disorder might be incriminating, but it is only circumstantial evidence, not a smoking gun.

Establishing causality in medicine requires more detective work.

Animal models can strengthen the case. Remember that the hippocampus in mice and men is nearly identical, down to the protein content of each of the hippocampal regions, so we can model in mice what we find in people. In a series of recent studies, when the proteins found in Alzheimer's disease were selectively manipulated to malfunction in mice, the entorhinal cortex was differentially affected, causing pathological forgetting. They were also found to contribute to the formation of amyloid plaques and neurofibrillary tangles, even leading ultimately to neuronal cell death. When, however, proteins found in normal aging were selectively manipulated, pathological forgetting was again observed, but this time the dentate gyrus was differentially affected. In agreement with what happens to us as we age normally, these manipulations caused neuronal sickness, but not amyloid plaques, neurofibrillary tangles, or cell death.

Biomedical detectives can also utilize genetics. In fact, recent genetic sleuthing has implicated genes that are associated with the proteins implicated in aging and that accelerate the slope of age-related memory decline. Other genetic investigations have isolated other genetic glitches, those associated with the proteins implicated in Alzheimer's disease, which increase the risk of getting the disease.

At this point, the detective work is nearing completion, and there is a convergence of incriminating evidence to bring the protein suspects to trial—which is to say a clinical trial, the only way to prove beyond a reasonable doubt that a suspected protein causes a disease. Developing safe interventions that correct a pathological protein is not easy. The good news is that many labs are tackling this problem, and so is the pharmaceutical industry. In the past few years, safe interventions have already been developed and have been found to correct the pathological proteins in animal models.

If Karl were alive to hear this latest update, I know I would have to once again ask my beloved and much-missed patient for patience as we continue to work to identify the causes of these two pathologies beyond a reasonable doubt—and thus the cures. That report would surely frustrate him, and perhaps does you as well. Believe me, the field is moving as fast as it can. There's no better way, perhaps, to end a book on normal forgetting than with a hopeful new beginning for pathological late-life forgetting. Stay tuned.

ACKNOWLEDGMENTS

Little did I know—ignorance, not hubris, I swear—that learning to write popular science was like learning to play a new instrument. I thank my editor at Crown, the unflappable Gillian Blake, for her master classes and patience in teaching me the mechanics, and assistant editor Caroline Wray for the remedial tutorials. I thank my wife, the pitch-perfect Alexis England, for the hours of listening and for her critical tuning, and my friend Sue Halpern, whose writing talents I now appreciate more than ever, for her words of encouragement. Finally, special thanks to the wonderful Alexandra Penney for introducing me to Gillian, and to my indomitable agent, Alice Martell.

NOTES

PROLOGUE

5 **Recent research in neurobiology** See, for example, Davis, R. L., and Y. Zhong, "The Biology of Forgetting—A Perspective." *Neuron,* 2017. 95(3): pp. 490–503; Richards, B. A., and P. W. Frankland, "The Persistence and Transience of Memory." *Neuron,* 2017. 94(6): pp. 1071–1084.

6 **When formally tested** Parker, E. S., L. Cahill, and J. L. McGaugh, "A Case of Unusual Autobiographical Remembering." *Neurocase,* 2006. 12(1): pp. 35–49.

6 **Jorge Luis Borges's short story** Borges, J., *Ficciones*. 1944, Buenos Aires: Grove Press.

ONE: TO REMEMBER, TO FORGET

20 **For more on this fascinating condition** Sacks, O., *The Man Who Mistook His Wife for a Hat*. 1985, London: Gerald Duckworth.

23 **Nevertheless, his legacy lives on** Augustinack, J. C., et al.,

"H.M.'s Contributions to Neuroscience: A Review and Autopsy Studies." *Hippocampus,* 2014. 24(11): pp. 1267–1268.

31 **So, more than a singular brain structure** Small, S. A., et al., "A Pathophysiological Framework of Hippocampal Dysfunction in Ageing and Disease." *Nature Reviews Neuroscience,* 2011. 12(10): pp. 585–601.

34 **My lab and others have been investigating** Brickman, A. M., et al., "Enhancing Dentate Gyrus Function with Dietary Flavanols Improves Cognition in Older Adults." *Nature Neuroscience,* 2014. 17(12): pp. 1798–1803; Anguera, J. A., et al., "Video Game Training Enhances Cognitive Control in Older Adults." *Nature,* 2013. 501(7465): pp. 97–101.

38 **New insight in the past few years** For example, Davis and Zhong, "The Biology of Forgetting"; Richards and Frankland, "The Persistence and Transience of Memory."

TWO: QUIET MINDS

42 **Kanner, who would become known as** Kanner, L., "The Conception of Wholes and Parts in Early Infantile Autism." *American Journal of Psychiatry,* 1951. 108(1): pp. 23–26; Kanner, L., "Autistic Disturbances of Affective Contact." *Nervous Child,* 1943. 2: pp. 217–240.

44 **we depend on our normal forgetting** Davis and Zhong, "The Biology of Forgetting"; Richards and Frankland, "The Persistence and Transience of Memory."

46 **like sculpting with marble** Migues, P. V., et al., "Blocking Synaptic Removal of GluA2-Containing AMPA Receptors Prevents the Natural Forgetting of Long-Term Memories." *Journal of Neuroscience,* 2016. 36(12): pp. 3481–3494; Dong, T., et al., "Inability to Activate Rac1-Dependent Forgetting Contributes to Behavioral Inflexibility in Mutants of Multiple Autism-Risk Genes." *Proceedings of the National Academy of Sciences of the United States of America,* 2016. 113(27): pp. 7644–7649.

50 **it seems to defy the idea** Khundrakpam, B. S., et al., "Cortical Thickness Abnormalities in Autism Spectrum Disorders Through

Late Childhood, Adolescence, and Adulthood: A Large-Scale MRI Study." *Cerebral Cortex,* 2017. 27(3): pp. 1721–1731.

50 **the proteins expressed by nearly all the genes** Bourgeron, T., "From the Genetic Architecture to Synaptic Plasticity in Autism Spectrum Disorder." *Nature Reviews Neuroscience,* 2015. 16(9): pp. 551–563.

51 **On average and as a group** Dong et al., "Inability to Activate Rac1"; Bourgeron, "From the Genetic Architecture to Synaptic Plasticity"; Tang, G., et al., "Loss of mTOR-Dependent Macroautophagy Causes Autistic-Like Synaptic Pruning Deficits." *Neuron,* 2014. 83(5): pp. 1131–1143.

51 **Among the few neuroimaging studies** See, for example, Corrigan, N. M., et al., "Toward a Better Understanding of the Savant Brain." *Comprehensive Psychiatry,* 2012. 53(6): pp. 706–717; Wallace, G. L., F. Happe, and J. N. Giedd, "A Case Study of a Multiply Talented Savant with an Autism Spectrum Disorder: Neuropsychological Functioning and Brain Morphometry." *Philosophical Transactions of the Royal Society B,* 2009. 364(1522): pp. 1425–1432.

52 **people with autism typically perform less well** Cooper, R. A., et al., "Reduced Hippocampal Functional Connectivity During Episodic Memory Retrieval in Autism." *Cerebral Cortex,* 2017. 27(2): pp. 888–902.

53 **When genes are manipulated to express** Dong et al., "Inability to Activate Rac1."

56 **The most successful computer algorithms** Masi, I., et al., "Deep Face Recognition: A Survey." IEEE Xplore, 2019.

58 **In computer science, this type of forgetting** Srivastava, N., et al., "Dropout: A Simple Way to Prevent Neural Networks from Overfitting." *Journal of Machine Learning Research,* 2014. 15: pp. 1929–1958.

59 **Nevertheless, the vast majority of studies confirm that sensory processing in autism** Behrmann, M., C. Thomas, and K. Humphreys, "Seeing It Differently: Visual Processing in Autism." *Trends in Cognitive Sciences,* 2006. 10(6): pp. 258–264.

60 **One of the most elegant studies** Pavlova, M. A., et al., "Social

Cognition in Autism: Face Tuning." *Scientific Reports,* 2017. 7(1): p. 2734.

61 **Another, older study** Frith, U., and B. Hermelin, "The Role of Visual and Motor Cues for Normal, Subnormal and Autistic Children." *Journal of Child Psychology and Psychiatry,* 1969. 10(3): pp. 153–163.

61 **Some psychologists have extended** Happe, F., "Central Coherence and Theory of Mind in Autism: Reading Homographs in Context." *British Journal of Developmental Psychology,* 1997. 15: pp. 10–12.

63 **Philosophers can debate** Rorty, R., *Philosophy and the Mirror of Nature*. 1979, Princeton, N.J.: Princeton University Press.

THREE: LIBERATED MINDS

73 **If factual information is processed** LaBar, K. S., and R. Cabeza, "Cognitive Neuroscience of Emotional Memory." *Nature Reviews Neuroscience,* 2006. 7(1): pp. 54–64.

76 **Recent functional imaging studies** Etkin, A., and T. D. Wager, "Functional Neuroimaging of Anxiety: A Meta-analysis of Emotional Processing in PTSD, Social Anxiety Disorder, and Specific Phobia." *American Journal of Psychiatry,* 2007. 164(10): pp. 1476–1488; Liberzon, I., and C. S. Sripada, "The Functional Neuroanatomy of PTSD: A Critical Review." *Progress in Brain Research,* 2008. 167: pp. 151–169.

77 **The general approach in treating PTSD** Etkin, A., et al., "Toward a Neurobiology of Psychotherapy: Basic Science and Clinical Applications." *Journal of Neuropsychiatry and Clinical Neurosciences,* 2005. 17(2): pp. 145–158.

78 **Currently, researchers are testing** Sessa, B., and D. Nutt, "Making a Medicine out of MDMA." *British Journal of Psychiatry,* 2015. 206(1): pp. 4–6.

78 **While there are no specific receptors for CBD** Piomelli, D., "The Molecular Logic of Endocannabinoid Signalling." *Nature Reviews Neuroscience,* 2003. 4(11): pp. 873–884; Bhattacharyya, S., et al., "Opposite Effects of Delta-9-Tetrahydrocannabinol and

Cannabidiol on Human Brain Function and Psychopathology." *Neuropsychopharmacology,* 2010. 35(3): pp. 764–774.

79 **He explained how the skits** Besser, A., et al., "Humor and Trauma-Related Psychopathology Among Survivors of Terror Attacks and Their Spouses." *Psychiatry: Interpersonal and Biological Processes,* 2015. 78(4): pp. 341–353.

79 **Most important in Yuval's view** Charuvastra, A., and M. Cloitre, "Social Bonds and Posttraumatic Stress Disorder." *Annual Review of Psychology,* 2008. 59: pp. 301–328.

FOUR: FEARLESS MINDS

86 **They retain childhood's playfulness** de Waal, F. B. M., *Peacemaking Among Primates.* 1989, Cambridge, Mass.: Harvard University Press, p. xi.

87 **In 2012, researchers accomplished** Rilling, J. K., et al., "Differences Between Chimpanzees and Bonobos in Neural Systems Supporting Social Cognition." *Social Cognitive and Affective Neuroscience,* 2012. 7(4): pp. 369–379; Issa, H. A., et al., "Comparison of Bonobo and Chimpanzee Brain Microstructure Reveals Differences in Socio-emotional Circuits." *Brain Structure and Function,* 2019. 224(1): pp. 239–251.

88 **These studies, crucial for mapping** Blair, R. J., "The Amygdala and Ventromedial Prefrontal Cortex in Morality and Psychopathy." *Trends in Cognitive Sciences,* 2007. 11(9): pp. 387–392.

90 **By stringing together a sequence** Cannon, W., "The Movements of the Stomach Studied by Means of the Roentegen Rays." *American Journal of Physiology,* 1896: pp. 360–381.

90 **He set out to study emotions** Cannon, W., *Bodily Changes in Pain, Hunger, Fear and Rage: An Account of Recent Researches into the Function of Emotional Excitement.* 1915, New York: D. Appleton & Company.

91 **Cannon's genius was in engineering** Cannon, W., and D. de la Paz, "Emotional Stimulation of Adrenal Secretion." *American Journal of Physiology,* 1911. 28(1): pp. 60–74.

93 **By the 1970s, it was clear** Swanson, L. W., and G. D. Petrovich, "What Is the Amygdala?" *Trends in Neurosciences,* 1998. 21(8): pp. 323–331.

94 **Scientists began making headway** LeDoux, J. E., "Emotion Circuits in the Brain." *Annual Review of Neuroscience,* 2000. 23: pp. 155–184.

95 **The latest study** Keifer, O. P., Jr., et al., "The Physiology of Fear: Reconceptualizing the Role of the Central Amygdala in Fear Learning." *Physiology (Bethesda, Md.),* 2015. 30(5): pp. 389–401.

97 **Evolutionary biologists have convincingly argued** Hare, B., V. Wobber, and R. Wrangham, "The Self-Domestication Hypothesis: Evolution of Bonobo Psychology Is Due to Selection Against Aggression." *Animal Behaviour,* 2012. 83(3): pp. 573–585.

97 **In a long-term study** Trut, L., "Early Canid Domestication: The Farm-Fox Experiment." *American Scientist,* 1999. 87: pp. 160–169.

100 **It turns out that many drugs** Roberto, M., et al., "Ethanol Increases GABAergic Transmission at Both Pre- and Postsynaptic Sites in Rat Central Amygdala Neurons." *Proceedings of the National Academy of Sciences of the United States of America,* 2003. 100(4): pp. 2053–2058.

100 **If you've tried the recreational drug MDMA** Carhart-Harris, R. L., et al., "The Effects of Acutely Administered 3,4-Methylenedioxymethamphetamine on Spontaneous Brain Function in Healthy Volunteers Measured with Arterial Spin Labeling and Blood Oxygen Level-Dependent Resting State Functional Connectivity." *Biological Psychiatry,* 2015. 78(8): pp. 554–562.

101 **So here we are** Young, L. J., "Being Human: Love: Neuroscience Reveals All." *Nature,* 2009. 457(7226): p. 148; Zeki, S., "The Neurobiology of Love." *FEBS Letters,* 2007. 581(14): pp. 2575–2579.

101 **Clues to the true purpose** Jurek, B., and I. D. Neumann, "The Oxytocin Receptor: From Intracellular Signaling to Behavior." *Physiological Reviews,* 2018. 98(3): pp. 1805–1908; Maroun, M., and S. Wagner, "Oxytocin and Memory of Emotional Stimuli: Some Dance to Remember, Some Dance to Forget." *Biological*

Psychiatry, 2016. 79(3): pp. 203–212; Geng, Y., et al., "Oxytocin Enhancement of Emotional Empathy: Generalization Across Cultures and Effects on Amygdala Activity." *Frontiers in Neuroscience,* 2018. 12: p. 512.

103 **when dogs and humans gaze** Nagasawa, M., et al., "Social Evolution. Oxytocin-Gaze Positive Loop and the Coevolution of Human-Dog Bonds." *Science,* 2015. 348(6232): pp. 333–336.

FIVE: LIGHTENING MINDS

109 **One telling anecdote** de Kooning, W., et al., *Willem de Kooning: The Late Paintings, the 1980s.* 1st ed. 1995, San Francisco: San Francisco Museum of Modern Art.

118 **Jasper did offer glimpses** Orton, F., *Figuring Jasper Johns.* 1994, London: Reaktion Books.

118 **Dreaming is known to be** Ritter, S. M., and A. Dijksterhuis, "Creativity—The Unconscious Foundations of the Incubation Period." *Frontiers in Human Neuroscience,* 2014. 8: p. 215.

120 **"We dream in order to forget"** Crick, F., and G. Mitchison, "The Function of Dream Sleep." *Nature,* 1983. 304(5922): pp. 111–114.

121 **they might even be deranged** Waters, F., et al., "Severe Sleep Deprivation Causes Hallucinations and a Gradual Progression Toward Psychosis with Increasing Time Awake." *Frontiers in Psychiatry,* 2018. 9: p. 303.

122 **In 2017, using powerful new microscopes** de Vivo, L., et al., "Ultrastructural Evidence for Synaptic Scaling Across the Wake/Sleep Cycle." *Science,* 2017. 355(6324): pp. 507–510; Diering, G. H., et al., "Homer1a Drives Homeostatic Scaling-Down of Excitatory Synapses During Sleep." *Science,* 2017. 355(6324): pp. 511–515; Poe, G. R., "Sleep Is for Forgetting." *Journal of Neuroscience,* 2017. 37(3): pp. 464–473.

122 **To paraphrase one of Crick's former students** Tononi, G., and C. Cirelli, "Sleep and the Price of Plasticity: From Synaptic and Cellular Homeostasis to Memory Consolidation and Integration." *Neuron,* 2014. 81(1): pp. 12–34.

123 **The behavioral consequences** Waters, "Severe Sleep Deprivation."

124 **Psychologists have pored over** Ghiselin, B., ed., *The Creative Process: Reflection on Invention in the Arts and Sciences*. 1985, Berkeley: University of California Press.

124 **Psychologists set out to devise** Mednick, S. A., "The Associative Basis of the Creative Process." *Psychological Review,* 1962. 69: pp. 220–232.

124 **for more examples** Bowden, E. M., and M. Jung-Beeman, "Normative Data for 144 Compound Remote Associate Problems." *Behavior Research Methods, Instruments, and Computers,* 2003. 35(4): pp. 634–639.

Remote Associate Items	**Solutions**
cottage/swiss/cake	cheese
cream/skate/water	ice
loser/throat/spot	sore
show/life/row	boat
night/wrist/stop	watch
duck/fold/dollar	bill
rocking/wheel/high	chair
dew/comb/bee	honey
fountain/baking/pop	soda
preserve/ranger/tropical	forest
aid/rubber/wagon	band
flake/mobile/cone	snow
cracker/fly/fighter	fire
safety/cushion/point	pin
cane/daddy/plum	sugar
dream/break/light	day
fish/mine/rush	gold
political/surprise/line	party
measure/worm/video	tape
high/district/house	school/court
sense/courtesy/place	common
worm/shelf/end	book
piece/mind/dating	game
flower/friend/scout	girl
river/note/account	bank
print/berry/bird	blue

Remote Associate Items	Solutions
pie/luck/belly	pot
date/alley/fold	blind
opera/hand/dish	soap
cadet/capsule/ship	space
fur/rack/tail	coat
stick/maker/point	match
hound/pressure/shot	blood

125 **Evidence that forgetting is beneficial** Storm, B. C., and T. N. Patel, "Forgetting as a Consequence and Enabler of Creative Thinking." *Journal of Experimental Psychology: Learning, Memory, and Cognition,* 2014. 40(6): pp. 1594–1609.

126 **other evidence** Ritter and Dijksterhuis, "Creativity."

SIX: HUMBLE MINDS

131 **we identified one group of nutrients** Brickman, "Enhancing Dentate Gyrus Function."

132 **I promised to send him** Barral, S., et al., "Genetic Variants in a 'cAMP Element Binding Protein' (CREB)–Dependent Histone Acetylation Pathway Influence Memory Performance in Cognitively Healthy Elderly Individuals." *Neurobiology of Aging,* 2014. 35(12): pp. 2881e7–2881e10.

134 **This one is called working memory** Lara, A. H., and J. D. Wallis, "The Role of Prefrontal Cortex in Working Memory: A Mini Review." *Frontiers in Systems Neuroscience,* 2015. 9: p. 173.

137 **Some, for reasons we still don't fully understand** Cosentino, S., et al., "Objective Metamemory Testing Captures Awareness of Deficit in Alzheimer's Disease." *Cortex,* 2007. 43(7): pp. 1004–1019.

138 **He had recently read** Schei, E., A. Fuks, and J. D. Boudreau, "Reflection in Medical Education: Intellectual Humility, Discovery, and Know-How." *Medicine, Health Care and Philosophy,* 2019. 22(2): pp. 167–178.

142 **"Judgment Under Uncertainty"** Tversky, A., and D. Kahneman, "Judgment Under Uncertainty: Heuristics and Biases." *Science,* 1974. 185(4157): pp. 1124–1131.

153 **At least this was the general assumption** Wimmer, G. E., and D. Shohamy, "Preference by Association: How Memory Mechanisms in the Hippocampus Bias Decisions." *Science,* 2012. 338(6104): pp. 270–273.

153 **We now know that hippocampal function** Shadlen, M. N., and D. Shohamy, "Decision Making and Sequential Sampling from Memory." *Neuron,* 2016. 90(5): pp. 927–939.

156 **most studies published before** Toplak, M. E., R. F. West, and K. E. Stanovich, "The Cognitive Reflection Test as a Predictor of Performance on Heuristics-and-Biases Tasks." *Memory and Cognition,* 2011. 39(7): pp. 1275–1289.

SEVEN: COMMUNAL MINDS

171 **Moral philosophers sometimes contrast** Margalit, A., *The Ethics of Memory*. 2002, Cambridge, Mass.: Harvard University Press, p. xi.

172 **most psychologists and sociologists agree** Lichtenfeld, S., et al., "Forgive and Forget: Differences Between Decisional and Emotional Forgiveness." *PLOS One,* 2015. 10(5): p. e0125561.

176 **"Clara" is actually a composite** Anspach, C., "Medical Dissertation of Nostalgia by Johannes Hofer, 1688." *Bulletin of the Institute of the History of Medicine,* 1934. 2: pp. 376–391.

179 **For those patients, exposure therapy** Kushner, M. G., et al., "D-Cycloserine Augmented Exposure Therapy for Obsessive-Compulsive Disorder." *Biological Psychiatry,* 2007. 62(8): pp. 835–838.

181 **Every nation has its own form** Boym, S., *The Future of Nostalgia.* 2001, New York: Basic Books.

EPILOGUE: PATHOLOGICAL FORGETTING

188 **"Listening to the cry"** Ventura, H. O., "Giovanni Battista Morgagni and the Foundation of Modern Medicine." *Clinical Cardiology,* 2000. 23(10): pp. 792–794.

191 **In 2001, we published our hypothesis** Small, S. A., "Age-

Related Memory Decline: Current Concepts and Future Directions." *Archives of Neurology,* 2001. 58(3): pp. 360–364.

192 **Our technical innovation succeeded** Small et al., "A Pathophysiological Framework."

193 **In the lower left panel** Khan, U. A., et al., "Molecular Drivers and Cortical Spread of Lateral Entorhinal Cortex Dysfunction in Preclinical Alzheimer's Disease." *Nature Neuroscience,* 2014. 17(2): pp. 304–311.

193 **The lower right panel** Brickman, "Enhancing Dentate Gyrus Function."

195 **As we hypothesized, the two populations** Small, S. A., "Isolating Pathogenic Mechanisms Embedded Within the Hippocampal Circuit Through Regional Vulnerability." *Neuron,* 2014. 84(1): pp. 32–39.

196 **In a series of recent studies** Small, S. A., and Petsko, G. A., "Endosomal Recycling Reconciles the Amyloid Hypothesis." *Science Translational Medicine,* 2020.

197 **The good news is** Mecozzi, V. J., et al., "Pharmacological Chaperones Stabilize Retromer to Limit APP Processing." *Nature Chemical Biology,* 2014. 10(6): pp. 443–449.

INDEX

ABOUT THE AUTHOR

SCOTT A. SMALL is a physician specializing in aging and dementia and a professor of neurology and psychiatry at Columbia University, where he is the director of the Alzheimer's Disease Research Center. He has run a National Institutes of Health–funded laboratory for over 20 years and has published more than 140 studies on memory function and malfunction. His work has been covered by *The New York Times, The New Yorker,* and *Time,* and his insight into Alzheimer's disease recently led to the formation of Retromer Therapeutics, a new biotechnology company, which he co-founded. Raised in Israel, he now lives in New York City.

ABOUT THE TYPE

This book was set in Berkeley, a typeface designed by Tony Stan (1917–88) in the early 1980s. It was inspired by, and is a variation on, University of California Old Style, created in the late 1930s by Frederic William Goudy (1865–1947) for the exclusive use of the University of California at Berkeley. The present face, in fact, bears influences of a number of Goudy's fonts, including Kennerley, Goudy Old Style, and Deepdene. Berkeley is notable for both its legibility and its lightness.